The Heparins
Basic and Clinical Aspects

The Heparins
Basic and Clinical Aspects

David Green, MD, PhD
Northwestern University Feinberg School of Medicine,
Chicago, IL, United States

ACADEMIC PRESS

An imprint of Elsevier

Academic Press is an imprint of Elsevier
125 London Wall, London EC2Y 5AS, United Kingdom
525 B Street, Suite 1650, San Diego, CA 92101, United States
50 Hampshire Street, 5th Floor, Cambridge, MA 02139, United States
The Boulevard, Langford Lane, Kidlington, Oxford OX5 1GB, United Kingdom

Notices
Knowledge and best practice in this field are constantly changing. As new research and experience broaden
our understanding, changes in research methods, professional practices, or medical treatment may become
necessary.

Practitioners and researchers must always rely on their own experience and knowledge in evaluating and
using any information, methods, compounds, or experiments described herein. In using such information or
methods they should be mindful of their own safety and the safety of others, including parties for whom they
have a professional responsibility.

To the fullest extent of the law, neither the Publisher nor the authors, contributors, or editors, assume any
liability for any injury and/or damage to persons or property as a matter of products liability, negligence or
otherwise, or from any use or operation of any methods, products, instructions, or ideas contained in the
material herein.

British Library Cataloguing-in-Publication Data
A catalogue record for this book is available from the British Library

Library of Congress Cataloging-in-Publication Data
A catalog record for this book is available from the Library of Congress

ISBN: 978-0-12-818781-4

For Information on all Academic Press publications
visit our website at https://www.elsevier.com/books-and-journals

Publisher: Stacy Masucci
Acquisition Editor: Tari Broderick
Editorial Project Manager: Tracy Tufaga
Production Project Manager: Kiruthika Govindaraju
Cover Designer: Miles Hitchen

Typeset by MPS Limited, Chennai, India

Working together
to grow libraries in
developing countries

www.elsevier.com • www.bookaid.org

Contents

Author biography

David Green, MD, PhD, is professor emeritus in Medicine, Division of Hematology/Oncology, at Northwestern University Feinberg School of Medicine in Chicago, Illinois. He received his medical degree from Jefferson Medical College in Philadelphia, Pennsylvania and doctorate in Biochemistry from Northwestern University, Evanston, Illinois. He is a clinician-investigator and author of more than 300 published scientific articles. His recent books include *Linked by Blood: Hemophilia and AIDS*, a description of the AIDS epidemic in the early 1980s that ravaged the hemophilia community, and *Hemophilia and Von Willebrand Disease*, exploring the role of the factor VIII/Von Willebrand factor complex in hemostasis and thrombosis. He is a master of the American College of Physicians and a distinguished emeritus member of the American Society of Hematology.

Acknowledgment

The author expresses appreciation to his wife, Theodora, for her cogent advice, support, and encouragement, and thanks Juan Chediak, MD, for the Case History included in Chapter 5.

Introduction

Unfractionated heparin, and more recently low molecular weight heparins and fondaparinux, have provided dependable prophylaxis and treatment for most thrombotic disorders for more than 100 years. These anticoagulants are now being replaced by the new, more patient-friendly direct oral anticoagulants. So why is a book about heparins appropriate at this time? Because heparins will remain indispensable agents for managing thrombosis into the forseeable future. Unfractionated heparin is uniquely suited to prevent clots in extracorporeal circuits and devices-it is effective immediately and can be rapidly reversed by protamine sulfate. Eight low molecular weight heparins are approved for clinical use; they have replaced unfractionated heparin for most other clotting conditions because they are simpler to administer, do not require monitoring, and have fewer adverse effects; in addition, they are safe to use for thrombotic disorders in pregnancy. Lastly, heparins are prepared from porcine intestines, a cheap and widely accessible source, and have nearly universal availability. The goal of this book is to assist practitioners in using these remarkable anticoagulants safely and effectively.

The family of heparins consists of unfractionated heparin, low molecular weight heparins (LMWHs), a synthetic pentasaccharide (fondaparinux), and several non-anticoagulant heparins. Heparins are sulfated glycosaminoglycans with molecular weights varying according to the composition and length of their constituent saccharide chains. Unfractionated heparin as currently manufactured usually has 40–60 saccharide units, corresponding to an average molecular weight of 12,000 to 20,000 Da, LMWHs 15–30 saccharides (4500–9000 Da) and synthetic heparin is a pentasaccharide with a molecular weight of 1728 Da. Except for the pentasaccharide, they are chemically heterogeneous. Non-anticoagulant heparins have saccharide chains of various lengths and degrees of sulfation. Unfractionated heparin, LMWH, and pentasaccharide are used for the prevention and treatment of thrombotic disorders, while non-anticoagulant heparins are being investigated for their activity in a broad spectrum of clinical disorders (Chapter 2).

The discovery of heparin was completely fortuitous. At the beginning of the 20th Century, surgeons wanted a substance that would accelerate blood coagulation and decrease surgical bleeding. William H Howell (1860–1945), a professor of physiology at Johns Hopkins, had prepared procoagulant phospholipids from brain tissue, but their coagulant activity was lost with storage. When Jay McLean (1897–1957) came to his laboratory requesting a research project, Howell assigned him the task of preparing a pure coagulant from the crude brain extracts then available.

McLean came from a family of illustrious medical people. His father was a physician, but died when McLean was only 4 years old. He had an uncle who had been Professor of Surgery and Dean at the University of California Medical School, and a cousin who was a Professor of Anatomy at Johns Hopkins prior to moving to the University of California. McLean was raised in San Francisco but the earthquake and fire of 1906 destroyed the family home and the place of employment of his stepfather [1]. McLean enrolled at the University of California at Berkeley but had to leave after two years because of financial difficulties. He took jobs as a gold miner, scrubbed decks on ferry boats, served as a railway mail clerk, and after 15 months was able to return to the University to complete his third and fourth years of study. Because his goal was to attend Johns Hopkins Medical School, he worked for several months as an oil well driller to earn money for the trans-continental trip to Baltimore. Unfortunately, his application for admission to the medical school was denied; he was told that he was "not the kind of man Hopkins sought." Nevertheless, he reapplied after taking courses in organic chemistry and German. Fortunately, there was an unexpected vacancy and he was accepted as a sophomore student, but then had to delay his medical studies because did not have enough money for tuition. He applied for an interim research position in Professor Howell's laboratory while he accumulated the funds needed to complete his training as a physician and surgeon.

McLean observed that the lipid fractions prepared from the ether/alcohol extraction of brain tissue contained a variety of phospholipids in addition to the clotting factor he was seeking, and wondered whether a procoagulant free of contaminating phospholipids might be obtained from other organs [1]. He perused the German chemical literature and found methods for extracting phospholipids from heart and liver. Testing the fractions from these tissues at regular intervals showed that their coagulant activity waned, just as was the case with brain extracts. However, the loss of clotting activity was accompanied by a prolongation of the clotting time of the test serum-plasma mixture, especially when liver extracts were examined. Rather than abandon the extracts because they had lost their procoagulant activity, he continued the testing, becoming convinced that a strong anticoagulant was present in liver tissue. He presented his observations to Howell who was dubious until McLean added the aged liver extract to fresh cat blood and showed that it completely prevented clotting. Howell was now satisfied that they had discovered a unique anticoagulant, and they called the inhibitor "heparin" because it was derived from the liver. Subsequently, Howell obtained higher yields using aqueous rather than ether/alcohol extraction [2], and the water-soluble fraction he retrieved probably contained authentic heparin.

One of the most remarkable attributes of heparin is its potent anticoagulant activity. This is demonstrated by wetting the inside of a syringe with a tiny drop of liquid removed from a vial of heparin. Blood subsequently drawn into this syringe will not clot even though there is no visible trace of the anticoagulant. Furthermore, very small amounts of heparin injected into animals or people will enhance the antithrombotic activity of the blood. When a constant infusion of only 1 IU/kg/hour was given intravenously for 3−5 days to patients undergoing surgery, the frequency of deep vein thrombosis and pulmonary embolism declined from 22% to 4% ($P < 0.01$) [3]. This potent medication has now been used for decades in the prevention and treatment of a variety of thrombotic disorders.

Initially, heparin was prepared by investigators interested in studying its characteristics and uses, but when it became an indispensable agent in medical practice, manufacturing was taken over by pharmaceutical firms and it was sold commercially. In the decades following World War II, thousands of people benefited from treatment with heparin and the anticoagulant was well-tolerated. However, during the latter part of 2007 and early 2008, severe hypersensitivity reactions were reported in dialysis patients; these untoward reactions were characterized by hypotension, facial swelling, urticaria, and tachycardia [4]. Symptoms appeared within minutes of initiation of a dialysis session; a total of 131 adverse events were reported from 21 facilities. Urgent epidemiologic investigations revealed that 97.7% of these events occurred shortly after the administration of heparin, and all of the heparin came from one manufacturer, Baxter Healthcare. The episodes subsided when Baxter withdrew the suspected lots of heparin.

Chemical analyses revealed that the Baxter heparin was heavily contaminated with over-sulfated chondroitin sulfate, a glycosaminoglycan found in cartilage [5]. Further study showed that this material directly activated the cellular kinin-kallikrein pathway to generate bradykinin; it also released anaphylatoxins derived from complement proteins, and induced hypotension when injected into swine. How did heparin, a trusted and widely used pharmaceutical produced by a major drug manufacturer, become contaminated with chondroitin sulfate?

At the time of these events, Baxter Healthcare was procuring its heparin from China [6]. The drug was extracted locally from porcine intestines, which were boiled and soaked in a resin bath. The crude heparin was shipped to plants in the U.S. and other countries for further processing. Chondroitin sulfate is not found in the intestines but is located in the cartilage of porcine snouts and ears. These animal parts of little commercial value might have been used as cheap stand-ins for mucosal heparin, and the extracted material added to the crude product to increase profitability [7]. It is estimated that use of the contaminated heparin resulted in at least 150 deaths in the U.S. alone [8].

This catastrophic event revived interest in heparin, at that time a 90-year old drug. In addition to examining its production and pharmacology, clinicians and others became motivated to discover and develop anticoagulants that were free of extraneous material, did not require monitoring, and were effective by subcutaneous injection or orally. In the ensuing years, low molecular weight heparin (LMWH), synthetic pentasaccharides, and direct oral anticoagulants were developed, underwent clinical trials, and entered clinical practice.

Bleeding is the most common adverse effect of heparin, and is usually from the gastrointestinal tract or other mucous membranes; the skin, muscles, and retroperitoneal space are other sites often affected. Major hemorrhages are relatively infrequent if dosing is adjusted for patient weight, age, and other factors such as thrombocytopenia and impairment of liver and renal function. While bleeding is not uncommon in patients receiving daily doses of heparins, single exposures to even large doses are well-tolerated; one expert recommended an initial bolus of 15,000 to 20,000 U of intravenous unfractionated heparin for patients with acute pulmonary emboli [9]. A few case reports also suggest that brief exposure to high doses of heparin are relatively safe. A nurse suspected of factitious heparin administration had coagulation studies that were consistent with the injection of a large amount of the anticoagulant; on examination, she had only a single ecchymosis [10]. Another patient who injected herself subcutaneously with 20 18,000-unit vials of the LMWH, dalteparin, only had a hematoma at the injection site; 80 hours were needed for the anti-Xa activity to return to normal [11].

Heparin-induced thrombocytopenia (HIT) is another serious adverse reaction to the anticoagulant. HIT can occur without warning after 5 days of heparin exposure; often, the first indication of the disorder might be a pulmonary embolus, myocardial infarction, or stroke. Because the risk of HIT is approximately 10-fold lower with LMWH than with unfractionated heparin, investigators conducted a 10-year study to determine whether avoiding the latter by using only LMWH would reduce the incidence of HIT and lower hospital costs [12]. The outcome was a 79% reduction in HIT cases and an 83% decrease in average estimated costs of HIT care per year. While HIT is infrequent with LMWH (0.9/10,000), the synthetic heparin, fondaparinux, virtually never induces HIT and has been used for the long-term management of HIT patients.

Is it possible to completely eliminate unfractionated heparin from medical practice, substituting LMWH, fondaparinux, or direct oral anticoagulants? At present, unfractionated heparin is the preferred anticoagulant for preventing thrombus formation in mechanical circulatory support devices for cardiopulmonary bypass, extracorporeal membrane oxygenation (ECMO), the total artificial heart, and venous access [13]. Alternative short-acting anticoagulants, such as bivalirudin or argatroban, are selected only for patients unable to tolerate heparin, and their inhibitory activity is not readily reversible. Unfractionated heparin, on the other hand, has the advantages of efficacy, reversibility with protamine, and low cost, making it difficult to replace.

Since the 2012 publication of *Heparin-A Century of Progress* [14], heparin regimens for the prophylaxis and treatment of thrombotic disease have improved, non-anticoagulant heparins with anti-inflammatory and antineoplastic activity have been described, and there have been advances in the recognition and management of heparin's adverse effects, particularly heparin-induced thrombocytopenia. The five chapters in this book review this progress and provide evidence-based recommendations for the use of heparins in the clinical management of thrombotic disorders. The expanded information about heparin's properties and applications contained in this book will be of benefit to clinicians and their patients.

References

[1] McLean J. The discovery of heparin. Circulation 1959;19:75−8.

[2] Howell WH. The purification of heparin and its presence in blood. Am J Physiol 1925;71:553−62.

[3] Negus D, Friedgood A, Cox SJ, Peel ALG, Wells BW. Ultra-low dose intravenous heparin in the prevention of postoperative deep-vein thrombosis. The Lancet 1980;i:891−4.

[4] Blossom DB, Kallen AJ, Patel PR, Elward A, Robinson L, Gao G, et al. Outbreak of adverse reactions associated with contaminated heparin. N Engl J Med 2008;359:2674−84.

[5] Kishimoto TK, Viswanathan K, Ganguly T, Elankumaran S, Smith S, Pelzer K, et al. Contaminated heparin associated with adverse clinical events and activation of the contact system. N Engl J Med 2008;358:2457−67.

[6] Fairclough G, Burton TM. The heparin trail. China's role in supply of drug is under fire. Wall St J (East Ed) 2008.

[7] Avorn J. Coagulation and adulteration-building on science and policy lessons from 1905. N Engl J Med 2008;358:2429−31.

[8] http://www.fda.gov/Drugs/DrugSafety/PostmarketDrugSafetyInformationforPatientsandPro viders/ucm112669.htm.

[9] Moser KM, Fedullo PF. Venous thromboembolism three simple decisions (Part 2). Chest 1983;83:256−60.

[10] Schmaier AH, Carabello J-A, Day HJ, Barry WE. Factitious heparin administration. Ann Intern Med 1981;95:592−3.

[11] Hasan K, Lazo-Langner A, Acedillo R, Zeller M, Hackam DG. Anticoagulant response after dalteparin overdose. J Thromb Haemost 2010;8:2321−3.

[12] McGowan KE, Makari J, Diamantouros A, Bucci C, Rempel P, Selby R, et al. Reducing the hospital burden of heparin-induced thrombocytopenia: impact of an avoid-heparin program. Blood 2016;127:1954−9.

[13] Kreuziger LB, Massicotte MP. Adult and pediatric mechanical circulation: a guide for the hematologist. Hematology Am Soc Hematol Educ Program 2018;2018:507−15.

[14] Lever R, Mulloy B, Page CP, editors. Heparin-a century of progress. Handbook of experimental pharmacology, vol. 207. Berlin Heidelberg: Springer-Verlag; 2012. 451 pp.

Part I

Historical Development and Properties

Chapter 1

Anticoagulant heparins

Discovery

Anticoagulant activity was detected by Jay McLean "incidentally" while he was attempting to purify cephalin from canine liver. McLean (1890−1957) had enrolled at Johns Hopkins School of Medicine as a second year medical student in 1915 and shortly thereafter began working in the laboratory of William Henry Howell (1860−1945), who assigned him the cephalin project. In the publication he prepared describing his results, he stated that the phospholipid he extracted from liver "has no thromboplastin action and in fact shows a marked power to inhibit the coagulation." [1] Later in 1916, McLean joined the Department of Research Medicine at the University of Pennsylvania, but Howell continued to study this anticoagulant activity, to which he and Emmett Holt gave the name, "heparin", because of its origin from the liver [2]. They observed that heparin was water-soluble when separated from cephalin, and a 0.1% solution prevented the clotting of shed blood.

As early as 1905, Morawitz [3] had proposed that blood coagulation was a two-step process: first, injured tissues released a factor, termed thromboplastin, which converted prothrombin to thrombin, and second, thrombin converted fibrinogen to fibrin. He also recognized that calcium was required for thromboplastin to be active. While this scheme could account for the clotting of shed blood, it did not explain why circulating blood remained fluid. In response, the studies of Howell and Holt [2] suggested that "heparin and pro-anti-thrombin…together fulfill the function of safeguarding the fluidity of the blood." They noted that the conversion of pro-antithrombin to antithrombin might protect against small amounts of thrombin, and that heparin functioned as a specific activator of pro-antithrombin and prevented the conversion of prothrombin to thrombin. Howell continued his anticoagulant studies, and in 1925 reported the isolation of a more reliable heparin that was water-soluble, free of protein, and prevented the clotting of 5 mL of blood in concentrations of 1 mg or less [4].

In 1928, Charles H. Best organized a group to examine the chemistry and physiology of heparin [5]. Their chemical studies were conducted at the Connaught Laboratories of the University of Toronto, the site where they produced insulin in the early 1920s. Best recruited David Scott and Arthur

The Heparins. DOI: https://doi.org/10.1016/B978-0-12-818781-4.00001-7
3

Charles, who reported that heparin could be isolated from beef lung, a much more accessible tissue, and following trypsin digestion could be crystallized as the barium salt [6]. This material had a potency of 100 arbitrary units/mg and they examined its effect on thrombi induced in canine veins. Removal of these veins several days after the injection of heparin showed complete resolution of thrombi and intimal healing.

Structure

Although producing heparin from animal tissues was not difficult, defining its chemical structure was challenging. This task was undertaken by J. Erik Jorpes at the Karolinska Institute in Stockholm, Sweden. Jorpes (1894–1973) received his medical degree from the University of Helsinki in 1925 and joined the Department of Chemistry and Pharmacology at the Karolinska; he rose through the academic ranks to become Professor of Medical Chemistry in 1946 [7]. In 1928–29, he visited the laboratories of Howell and Best, and on his return to Sweden, he organized the production of insulin and heparin [8]. In 1935, Jorpes showed that the main carbohydrate constituents of heparin were uronic acid and hexosamine, present in a 1:1 ratio [9]. Subsequently, Jorpes and Bergstrom [10] reported that the anticoagulant activity of heparin could be enhanced by increasing the extent of sulfation. However, a member of Best's group, Louis Jaques, observed that heparins derived from different species varied in anticoagulant potency despite having similar sulfur content, suggesting that species-specific factors contributed to the anticoagulant activity [11]. Jaques noted that the heparin polymer consisted of alternating sulfoglucosamine and hexuronic acid moieties joined by glycosidic linkages, and that these were present in amounts specific for each species [12]. Active chains generally had molecular weights in the 12,000–20,000 dalton range. He concluded that heparins "are mixtures of individual highly negatively charged chains that show a wide spectrum of specific reactions with biologically active proteins." The hexuronic acid was identified as L-iduronic acid by Cifonelli and Dorfman [13] in 1962; the structures of D-glucosamine and L-iduronic acid are shown in Fig. 1.1.

In 1936, Hjalmar Holmgren, a histology assistant in Jorpes laboratory, was tasked with determining the cell of origin of heparin. He used toluidine blue staining to determine that the anticoagulant was located in the metachromatic granules of mast cells. In fact, glycosaminoglycans (GAGs) such as heparin account for as much as 25% of the total organic content of rat mature mast cells [14]. Mast cells from human lung are estimated to contain heparin in a concentration of 2.4–7.8 µg/106 cells [15]. Tissue mast cells have a very characteristic appearance with metachromatic dyes, displaying extensive grapelike clusters of granules (Fig. 1.2). Heparin is synthesized in the Golgi compartment of the mast cells and attached to the serglycin core protein [16]. A critical enzyme for the modification of the nascent heparin

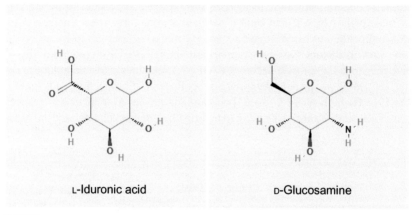

FIGURE 1.1   L-Iduronic acid and D-glucosamine. *Images from https://pubchem.ncbi.nlm.nih,gov.*

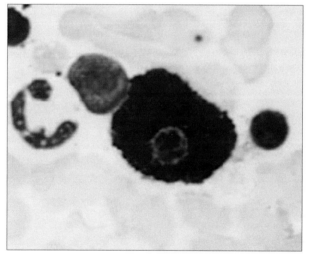

FIGURE 1.2   Mast cell in the bone marrow, Giemsa stain, 1000 ×. *Photomicrograph courtesy of Yi-Hua Chen, MD, Feinberg School of Medicine, Northwestern University.*

chains is N-acetylase/N-sulfatransferase 2 (NDST2), which is expressed in massive amounts in mast cells and is essential for the N-deacetylation and N-sulfation of the nascent polysaccharide chains. In the absence of this enzyme, mast cells are abnormal and lack heparin [17]. In contrast, heparan sulfate is made by cells throughout the body, and the key enzyme required for its synthesis is NDST1.

Glycosaminoglycans (GAGs) are linear polysaccharide chains consisting of repeating disaccharide groups. They are found in the extracellular matrix and bind to a core protein to form proteoglycans. The principal proteoglycans are hyaluronan and keratan, chondroitin, dermatan, and heparan sulfates. Heparan sulfate is located on the surfaces and basement membranes of most cells where it can bind a number of effectors. These include adhesion molecules, tyrosine kinase receptors, integrins, and complement receptors; growth factors such as vascular endothelial (VEGF), epidermal (EGF), and transforming growth factor-β (TGF-β); cytokines and chemokines; and the extracellular matrix components: collagens, fibronectin, and laminin [18,19]. Heparin differs from heparan sulfate in a number of respects (Fig. 1.3 and Table 1.1) [20]. Heparin has fewer chains and a small

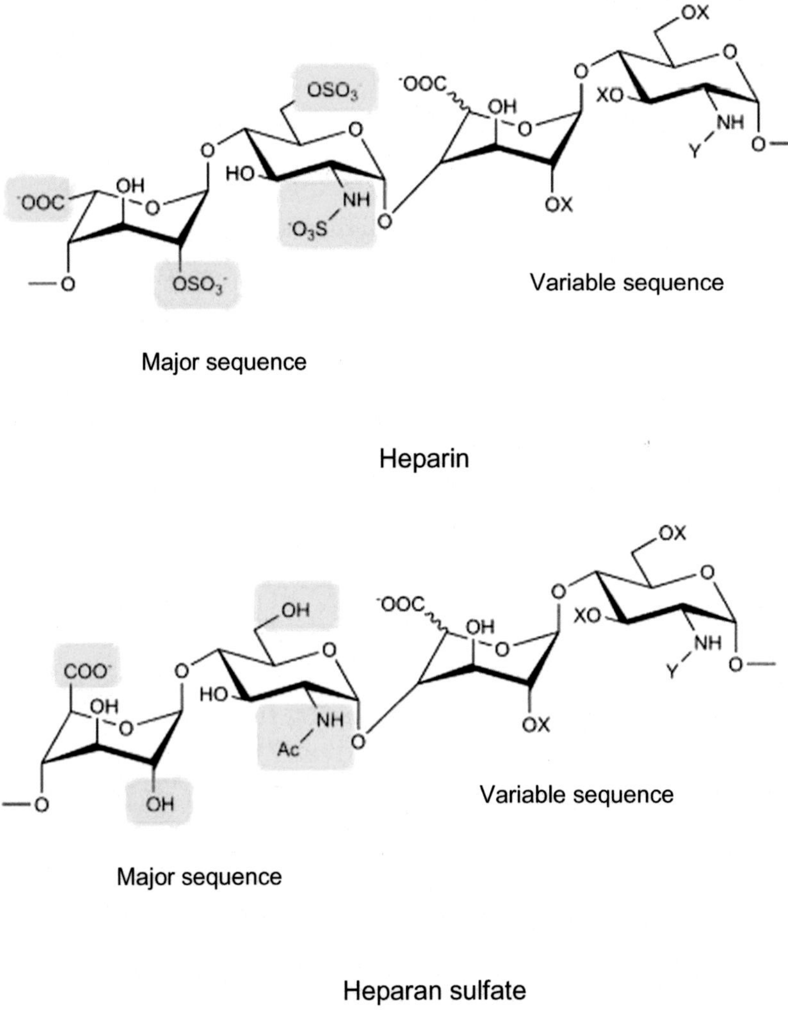

FIGURE 1.3 Heparin and Heparan. *From Noti C, Seeberger PH. Chemical approaches to define the structure-activity relationship of heparin-like glycosaminoglycans. Chem Biol 2005;12:731–56.*

TABLE 1.1 Comparison of heparin and heparan sulfate.

Property	Heparin	Heparan sulfate
Molecular mass (kDa)	~15	~30
Ratio of iduronic acid to glucuronic acid	>1	<1
Sulfate groups per disaccharide unit	3	1
N-Sulfated glucosamines (%)	80–90	30–60

TABLE 1.2 Low molecular weight heparins.

Product	Depolymerization method	Molecular weight (da)	Ratio anti-Xa/ anti-IIa
Bemiparin	Alkaline degradation	3600	8.0
Nadroparin	Deamination by nitrous acid	4300	3.3
Reviparin	Deamination by nitrous acid	4400	4.2
Enoxaparin	β-Cleavage by alkaline benzylation	4500	3.9
Parnaparin	Oxidation with Cu^+/H_2O_2	5000	2.3
Dalteparin	Deamination by nitrous acid	5000	2.4
Certoparin	Deamination by nitrous acid	5400	2.5
Tinzaparin	β-Cleavage by heparinase	6500	1.6

molecular mass (15 vs 30 kDa, and >80% of the glucosamine residues are N-sulfated, but the concentration of O-sulfates exceeds N-sulfates). In heparan, the numbers of N-sulfate and N-acetyl substituents are approximately equal and the number of O-sulfate groups can range from 20 to 75 per 100 disaccharide units [21]. Heparin also has a higher sulfate to disaccharide ratio and more iduronic acid [22]. In fact, heparin has the highest negative charge density and is the most acidic biologic macromolecule in the human body [23].

Low molecular weight heparins (LMWHs) are prepared by depolymerization of the parent compound using the variety of patented enzymatic and chemical procedures shown in Table 1.2. The average molecular weights of

LMWHs range from 3600 to 6600 da, and consist of 12—22 saccharide units; 15%—25% of the chains have the pentasaccharide binding site for antithrombin [24]. The size distribution of the chains is variable from product to product; for example, the smaller chains (<3000 da) comprise 3%—15% of dalteparin but about 20% of enoxaparin, whereas the larger chains (>8000 da) are present in 14%—26% of dalteparin but ≤15% of enoxaparin. Each product has distinctive properties and is considered unique for licensing purposes.

The synthetic pentasaccharide, fondaparinux, is a selective inhibitor of factor Xa (FXa) with a molecular weight of 1728 da. Four of its five saccharide units are sulfated, enhancing its affinity for antithrombin and its anti-Xa activity [25]. It has an elimination half-life of 17 hours. Fondaparinux requires a complex chemical synthesis involving some 50 steps. An analog, idraparinux, is an O-methylated, O-sulfated pentasaccharide with an even higher affinity for antithrombin and a prolonged half-life (4 days) [26]. The addition of a biotin moiety to idraparinux (idrabiotaparinux) enabled the rapid reversal of anticoagulant activity by infusion of avidin [27]. Idrabiotaparinux was compared with idraparinux in patients with symptomatic deep venous thrombosis and found to have a similar time course of FXa inhibition, safety, and efficacy [28].

Pharmacokinetics

Unfractionated heparin is generally given parenterally because it is poorly absorbed from mucosal surfaces. Following subcutaneous injection, absorption is dose-dependent, ranging from 5% to 50% with doses of 5,000 U to 25,000 U. In a clinical trial, the heparin level was only 0.01 U/mL 6 hours after a dose of 5,000 U, but it was 0.18 U/mL after a dose of 12,500 U [29]. The bioavailability of subcutaneously injected heparin is also affected by the particular preparation (calcium or sodium salt) [30] and the skinfold thickness [31].

Even when infused intravenously, the half-life of heparin is quite variable, depending on the extent of binding to plasma proteins such as lipoproteins, immunoglobulins, and fibrinogen as well as macrophages [12,32,33]. In addition, heparin is cleared by the hyaluronic acid receptor for endocytosis (HARE/stabilin-2), expressed by the sinusoidal endothelial cells of the liver and lymph nodes [34—36]. Heparin binds to HARE in clathrin-coated pits, and is endocytosed and degraded. The half-life can vary dose-dependently from 30 minutes with a dose of 25 U/kg, 56 minutes with a dose of 100 U/kg, and 152 minutes with 400 U/kg [37,38]. The disappearance of the anticoagulant activity follows nonlinear kinetics, but once all the cellular receptors become saturated, non-saturable heparin levels rise linearly with the dose administered [39].

LMWHs have a more predictable anticoagulant response than unfractio-nated heparin. Following subcutaneous injection of dalteparin, enoxaparin, or tinzaparin, bioavailability is 87%, 92%, or 90%, respectively, whereas the bioavailability of unfractionated heparin varies greatly with the dose injected. The disappearance of LMWH is linear and the half-life is twice as long as observed with an equivalent dose of unfractionated heparin. LMWHs bind less to plasma proteins and their cellular clearance is less efficient [40,41]. Despite more predictable pharmacokinetics, there is considerable inter-individual variability in coagulation assays, half-life, and clearance after a uniform dose of LMWH. For example, the anti-Xa activity of enoxaparin exceeds that of nadroparin by a factor of 1.48, and dalteparin by a factor of 2.28, and its elimination half-life is 4.1 hours compared to 3.7 for nadroparin and 2.8 for dalteparin [42]. Nevertheless, when dosed according to the manu-facturer's instructions, the clinical effectiveness of each LMWH appears to be similar and monitoring of blood levels is usually unnecessary.

The elimination of non-saturable heparin is by the kidney, where it is de-sulfated and excreted as uroheparin [43]. In patients with chronic renal dis-ease, post-infusion heparin levels are significantly higher than in patients with normal renal function (odds ratio 2.04; 95% CI, 1.27−3.27), and high levels are more persistent [44]. When the creatinine clearance is <60 mL/min, clotting times outside he therapeutic range are most often associated with heparin bolus doses that exceed 130 U/kg. LMWHs, especially the smaller chains, are mainly excreted by the kidneys and their clearance is strongly affected by renal disease; for example, after intravenous injection of 100 U of nadroparin, MW 4.3 kDa, the half-life of anti-FXa activity was 2.2 ± 0.4 hours in individuals with normal renal function, but 4.1 ± 0.6 hours in patients with chronic renal insufficiency [41]. Consequently, there is a ten-dency for the smaller LMWHs to accumulate in patients with impaired renal function if doses are not reduced [45]. A meta-analysis found that non-dialysis patients with a creatinine clearance of ≤ 30 ml/minute receiving standard therapeutic doses of enoxaparin, MW 4500 da, had elevated anti-Xa levels and an increased risk of major hemorrhage [46]. Although tinzaparin, MW 6500 da, does not appear to accumulate in older patients with reduced renal function [47], this LMWH was associated with a higher mortality than unfractionated heparin in a randomized trial evaluating treatments for acute deep vein thrombosis in the elderly [48].

Fondaparinux is a synthetic pentasaccharide that is given once daily sub-cutaneously and is 100% bioavailable. After a dose of 2.5 mg, peak levels are reached in two hours and distribution is within the intravascular compart-ment, with minimal binding to plasma proteins other than antithrombin. The drug is eliminated unchanged in the urine with an elimination half-life of 17−21 hours, but this is prolonged in individuals aged ≥ 75 [49]. Clearance is also decreased by approximately 30% in those weighing less than 50 kg. Because excretion is mainly by the kidney, the percent drug cleared is progressively reduced in patients with renal disease: 25% with a creatinine clearance (ml/minute) 50−80; 40% with 30−50; and 55% with <30. Therefore, the drug is contraindicated in patients with severe renal disease and should be used cautiously in those with any impairment of renal function.

Pharmacodynamics

In 1939 Brinkhous et al. [50] demonstrated that the inhibitory activity of heparin was absolutely dependent on a non-diffusible cofactor present in plasma and serum. This cofactor was identified as antithrombin III and now simply designated antithrombin, and it progressively inactivated thrombin; the rate of inactivation was greatly accelerated by heparin [51,52]. At a concentration of 10^{-8} to 10^{-7} mol/L, the rate of antithrombin/thrombin interaction was increased 10,000-fold, and antithrombin/FXa interaction 4500-fold [53]. Antithrombin is a serine protease inhibitor or serpin, a protein that presents a reactive loop mimicking a serine protease substrate. Proteases such as thrombin and FXa become trapped when they attempt to cleave this loop.

Antithrombin was isolated by Rosenberg and Damus in 1973, and was observed to form a 1:1 stoichiometric complex with thrombin [54]. They surmised that an arginine residue of antithrombin bound to the active site serine of thrombin, and lysyl residues probably served as the binding site for heparin. Heparin binding induced a conformational change in antithrombin that potentiated the inactivation of thrombin. Unfractionated heparin simultaneously binds to both antithrombin and thrombin (Fig. 1.4). Rosenberg observed that only one-third of the heparin molecules formed a stable complex with antithrombin. He investigated those heparin fractions that had high affinity for antithrombin, and suggested that a tetrasaccharide sequence represented the critical site responsible for heparin's anticoagulant activity [55]. However, Choay et al. [56] synthesized saccharides of various lengths (Fig. 1.5, sequence 2) and showed conclusively that the synthetic pentasaccharide had the strongest binding to antithrombin (Ka: 7×10^6/M) as well as enhanced inhibitory activity towards factor Xa. His demonstration that the pentasaccharide was the minimal sequence required for binding to antithrombin eventually led to the clinical development of the synthetic pentasaccharide, fondaparinux. A further advance was the recognition that a 3-O-sulfate group in the pentasaccharide is essential for the binding of antithrombin to heparin [57].

In 1973, methods were described for estimating heparin by measuring residual FXa activity [58,59]. These assays showed that LMWHs had less thrombin inhibition than unfractionated heparin but retained their anti-FXa activity (Fig. 1.4) [60]. Thrombin inhibition was decreased with heparin chains less than 30 saccharide units in length (MW 9000 da), and was completely lost when there were fewer than 12 saccharide units

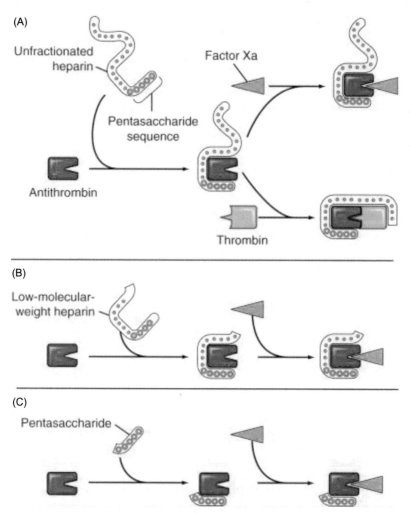

FIGURE 1.4 Mechanism of action of heparin, low-molecular-weight heparin, and fondapari-nux. All heparins with the pentasaccharide sequence bind to antithrombin, inducing a conforma-tional change that accelerates inactivation of FXa. Panel A. Unfractionated heparin binds both thrombin and antithrombin, enhancing thrombin inhibition. Panels B & C. Low-molecular-weight heparin chains with ≤ 12 saccharide units adjacent to the pentasaccharide sequence and fondaparinux are too small to enhance the inhibition of thrombin by antithrombin. *From Chapter 149 in Hoffman R, Benz EJ, Jr, Silberstein LE, Heslop HE, Weitz JI, Anastasi J, editors.* Hematology basic principles and practice, *66th ed. Canada: Saunders an imprint of Elsevier Inc; 2018, p. 2174.*

(MW 3600 da) adjacent to the pentasaccharide sequence [61]. Consequently, it was thought unlikely that smaller heparin fractions would have clinical utility as anticoagulants. This idea was refuted by the work of Edward Johnson and

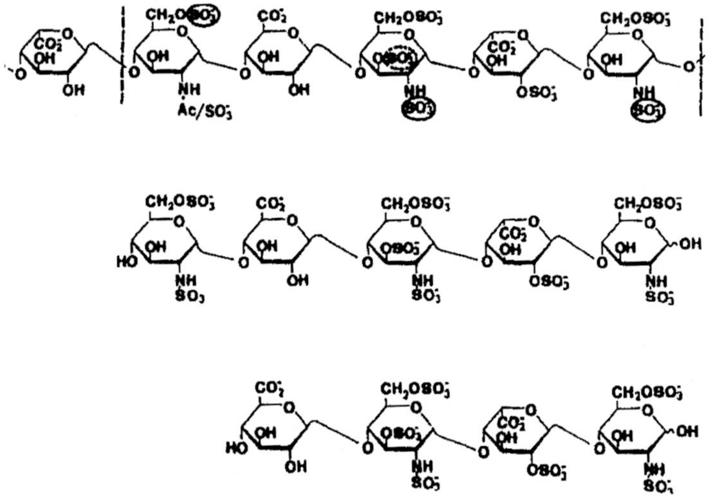

FIGURE 1.5 Heparin polysaccharides: (1) hexasaccharide; (2) pentasaccharide; (3) tetrasaccharide. *Reproduced from Choay J, Petitou M, Lormeau JC, Sinay P, Casu B., Gatti G. Structure-activity relationship in heparin: a synthetic pentasaccharide with high affinity for antithrombin III and eliciting high anti-factor Xa activity. Biochem Biophys Res Commun 1983;116:492–99.*

colleagues [62], who prepared heparin fractions varying in MW from 9000 to 22000 da. These fractions were injected subcutaneously in volunteers and the effect on coagulation measured using the anti-Xa assay. Unexpectedly, the lowest MW fraction had the most potentiating effect on the assay. This suggested that LMWH, although a weak inhibitor of thrombin, had potent anti-FXa activity and led to clinical trials testing the effectiveness of this heparin formulation for the prevention and treatment of thrombosis. Similarly, fondaparinux, while devoid of thrombin inhibiting activity, has potent anti-FXa activity [63].

More recent studies show that the conformational changes induced by the binding of the pentasaccharide to antithrombin culminate in the expulsion of the reactive loop and enhance its display to such proteases as thrombin and FXa [64]. Thrombin bound to heparin cleaves the loop sequence, is trapped by the covalent linkage, and becomes inactivated. On the other hand, FXa not only interacts with the reactive loop, but also binds to a specific exosite on antithrombin. The binding of the protease to the exosite enhances heparin's allosteric activation of antithrombin by supporting positive and dampening negative exosite interactions [65]. In 1977, Carlstrom et al. [66] observed that the interaction between antithrombin and thrombin decreases the affinity of heparin for antithrombin, enabling heparin to dissociate from the trimolecular complex and bind more free antithrombin molecules.

The molecular mechanism underlying the anticoagulant action of heparin has been extensively investigated by H.C. Hemker (1934-). Hemker received a medical degree in 1959 and a doctorate in biochemistry in 1962. He led a research group at the University of Leiden until 1975, when he moved to the University of Maastricht, where he built a major research group specializing in hemostasis and thrombosis. He was invited to be a Visiting Professor at the College de France in Paris and at Mount Sinai Medical School in New York, and has received multiple prizes and awards. These include Knight of the Dutch Lion and the Legion of Honor, as well as a Commander of the Order of the Palmes Academiques. He has authored more than 500 scientific publications and holds several patents, principally on devices and methods for quantitating thrombin generation.

Hemker recognized that the crucial function of heparin is to catalyze the irreversible inactivation of clotting enzymes by antithrombin. He and his staff showed that the capacity to inhibit coagulation primarily resides in the pentasaccharide and the 12 saccharide units at the non-reducing end of the heparin molecule [67]. Thrombin binds to the longer heparin chains and moves by unidimensional diffusion toward the bound antithrombin, now in a conformation capable of its rapid inactivation [68]. However, the binding of FXa is 11 times weaker than that of thrombin, and its inactivation is three times less efficient [69]. To study heparin's inhibition of thrombin generation, Hemker and colleagues developed a new tool called "the calibrated automated thrombogram (CAT)" [70]. This enabled them to monitor thrombin generation in clotting plasma and demonstrate that a ten-fold higher concentration of pentasaccharide is required for inhibition of the endogenous thrombin potential when using LMWH as compared with unfractionated heparin. Hemker has argued that the CAT might be a better measure of heparin's anti-thrombotic prowess because the anti-FXa test reflects heparin molecules with anti-thrombin activity as well as the "practically inert" molecules that have only anti-FXa activity [69]. For example, the inhibitory effects on thrombin generation are greater with the largest (i.e., tinzaparin) than the smallest (i.e., bemiparin) LMWHs when compared by anti-Xa activity, but are similar when compared by anti-thrombin activity [71].

Heparin's inhibition of coagulation is not limited to activation of antithrombin; even in the absence of antithrombin, heparin prolongs the activated partial thromboplastin time [72]. Intravenous infusions of heparin release tissue factor pathway inhibitor (TFPI) from the endothelium [73,74], and promote complex formation between thrombin and plasminogen activator inhibitor-1 (PAI-1), enhancing thrombin inhibition and indirectly accelerating fibrinolysis [75]. In addition, heparin does not affect the binding of tissue plasminogen activator to thrombi or otherwise interfere with thrombolysis [76]. Both the release of TFPI and the PAI-1 induced thrombin inhibition are less pronounced with low molecular weight heparin (LMWH) and are absent with fondaparinux [63,75,77]. Unfractionated heparin enhances the anticoagulant activity of activated protein C (APC); the generation of APC activity by LMWHs is correlated with their molecular weight [78]. Heparin also binds to monocytes and macrophages, and the cell-surface bound heparin retains the ability to accelerate the inactivation of thrombin by antithrombin [79]. It also enhances the inhibitory activity of antithrombin toward other coagulation proteases, such as FIXa and FXIa [80]. On the other hand, heparin increases the activation of FXI by thrombin-bound fibrin [81].

Heparin exhibits several interactions with platelets. In 1971, Eika observed that heparin aggregated washed platelets in the presence of calcium, and subsequently he reported that aggregation was due to release of platelet adenosine diphosphate (ADP) [82,83]. The mechanism that accounts for this platelet activation is the binding of heparin to platelet surface membrane glycoprotein (Gp) αIIbβ3, stimulating key signaling and cytoskeletal changes [84]. However, when heparin (100 U/kg) is given by intravenous injection to healthy volunteers, platelet function is inhibited; the bleeding time is prolonged and the platelet release reaction impaired [85]. Most of the heparin binds to antithrombin and this bound heparin is not absorbed by platelets [86]. Furthermore, heparin, antithrombin, and thrombin form a trimolecular complex that strongly limits the ability of thrombin to induce platelet aggregation [87], which might explain the prolonged bleeding time after heparin infusion. Heparin also inhibits platelet aggregation and release induced by neutrophil cathepsin G [88], and neutralizes the pro-thrombotic activity of platelet factor 4 [89]. On the other hand, heparin binds to the A1 domain of von Willebrand factor (VWF), at aminoacids 565-587 and 642-645 [90,91], potentiating the interaction of VWF with platelet GPIbα [92]. Heparin, at concentrations that inhibited coagulation, did not prevent the thrombin-induced platelet activation and vessel wall deposition that occur in response to severe arterial injury [93]. This outcome is not unexpected because higher anticoagulant concentrations are required to inhibit arterial than venous thrombosis. Decreased platelet deposition in a baboon model of arterial thrombosis was observed when sufficient unfractionated heparin or LMWH was given to raise plasma anti-Xa levels from 0.5 U/mL to 1−5 U/mL [94].

Clinical development

Investigators initiated studies of heparin preparations in human subjects as early as 1924, but their crude preparations elicited adverse reactions and were poorly tolerated. The crystalline barium salt prepared by Charles and Scott [6] in 1936 did not have untoward side effects and was enlisted for the first patient treatments in Canada. This method of heparin preparation was also used by Jorpes in Sweden, and clinicians in both countries initiated treatment of patients with thrombotic disorders.

Clinicians observed that the inhibition of coagulation by heparin was of brief duration. Chargaff and Olson [95] reasoned that the anticoagulant effect might be extended by protamine because this agent prolonged the action of insulin (protamine-zinc insulin). Unexpectedly, they found that protamine completely inhibited heparin activity, and showed in a dog model that prot-amine could be used to reverse heparin-induced anticoagulation. By identify-ing a safe and effective antidote to heparin, they contributed to the continued interest in the clinical development of the drug. Erwin Chargaff next turned his attention to the chemistry of nucleic acids, and became famous for his discovery that DNA is the primary constituent of the gene.

The first commercially manufactured heparin for the US market became available in 1939. It was Liquaemin (Roche/Organon), and its production by the Dutch company, Organon, was in response to the demand for large amounts of heparin to prevent clot formation in the artificial kidney under development in the Netherlands [69]. It was prepared from beef lung and required extreme conditions including heating at 95 °C, pH 2, and alcohol precipitation, which effectively sterilized the product. In the 1940s, decreased availability of beef lungs and ready access to porcine intestinal mucosa as a byproduct of sausage manufacture led to selection of the porcine tissue as the preferred source of heparin, completely replacing beef lung by the 1970s. It is now estimated that the worldwide production of crude hepa-rin is in excess of 110 tons per year [96].

Although heparan sulfate with anticoagulant activity was identified on the surface of cultured endothelial cells [97,98], only very small amounts (44 ± 14 nmol uronic acid/dl) circulate in the plasma, mostly in non-covalent complexes with plasma proteins [99]. Mucosal heparin was found to have more anticoagulant activity in most assays [100]. Anticoagulant potency var-ied not only by the species of origin, but also by manufacturing method, so that a milligram of one preparation might not be comparable to a milligram of another. When heparin first became available, clinicians gave amounts that totally prevented clotting, but when smaller, more nuanced effects on coagulation were desired, having accurate standardization of the drug became imperative [101]. In 1947, the World Health Organization (WHO) designated the reference heparin produced by Murray and Best as the 1st International Standard, and in 1973, replaced it with porcine mucosal heparin [102]. These Standards defined the International Unit (IU) of heparin. The United States Pharmacopeia (USP) developed its own Unit; one USP unit was defined as the minimal amount of heparin which, when added to citrated sheep plasma, prevented the clotting of 1 mL of plasma for 1 hour after the addition of 0.2 mL of a 1:100 solution of calcium chloride [103]. Mucosal heparin standards had a potency of 140 U/mg and beef lung standards, 120 U/mg. However, assays based on coagulation tests were relatively unin-formative about the relative efficacy of heparin preparations in controlling thrombosis [104]. More recently, the USP has replaced the clotting assay with a chromogenic anti-thrombin test, and in 2009, introduced a new potency reference standard calibrated relative to the WHO International Standard so that 1 USP unit is equivalent to 1 IU of heparin. The potency of the heparin sodium standard, calculated on a dried basis, is 180 USP Heparin Units per mg [105].

Standardization of LMWH has proved more difficult because at least eight different products are licensed globally and currently marketed; attempting to use unfractionated heparin as the reference standard for anti-Xa and anti-IIa assays led to high inter-laboratory variability and invalid assays due to non-parallelism [106]. Using a LMWH as the standard for the others gave parallel assays and closer agreement between laboratories; investigators selected the most representative LMWH as the reference standard. The potency of the LMWHs is estimated using chromogenic assays; [107] for example, dalteparin and enoxaparin have anti-Xa activities of 156 IU/mg and 100 IU/mg, respectively, with reference to the WHO First International Low Molecular Weight Reference Heparin Standard. In the past 20 years, Second and Third International Standards for Low Molecular Weight Heparins have been established; the batches now in use are assigned anti-Xa activities of 102 and 100 IU/vial and anti-IIa of 34 and 33 IU/vial [108,109]. The ratio of anti-Xa to anti-IIa is characteristic of each LMWH; the smaller ones have the least anti-IIa activity but all have an anti-Xa potency >70 IU/mg. Anti-Xa to anti-IIa ratios range from 8.0 for bemiparin to 1.6 for tinzaparin (Table 1.2).

Among the investigators studying heparin in the 1930s, surgeons in particular recognized its potential to treat thrombotic disorders. A tragic and frustrating surgical experience is the successful performance of a difficult operation with minimal blood loss, only to have the patient die shortly postoperatively of a massive pulmonary embolism. Clarence Crafoord, a Swedish thoracic surgeon, was well-aware of this complication, having performed several pulmonary embolectomies on postoperative patients. He recognized that these thromboses might be averted by systemic anticoagulant therapy, and when heparin became available in 1935 from the Jorpes laboratory, Crafoord and his colleagues began administering it to their postoperative and postpartum patients. No instances of thrombosis were recorded in 657 patients receiving at least 250 mg heparin daily for 5–10 days following operative procedures [110]. Jorpes also records that daily infusions of his heparin restored the visual acuity of a patient with acute central retinal vein thrombosis and resulted in clinical improvement in a patient with posterior inferior cerebellar artery occlusion. At roughly the same time, a Canadian surgeon, GDW Murray, and Best reported that heparin was effective in preventing pulmonary embolism in 315 postoperative patients; in addition, 28 patients with thrombophlebitis and seven with pulmonary embolism were successfully treated [5,111].

The clinical use of heparin for prophylaxis and treatment of thrombosis was stimulated by an increase in the occurrence of venous thromboembolism during World War II. Doctor Keith Simpson, a Lecturer in Forensic Medicine to Guy's Hospital, reported 24 autopsied cases of pulmonary embolism in 1940, as compared to only 4 in 1939 [112]. More than 80% of those dying were either in air-raid shelters or death occurred soon after leaving such facilities. In bomb shelters, people were relatively immobile for prolonged periods. Simpson described a 60 year old woman who sat in a deckchair continuously for 10 hours. In such chairs, the front edge of the seat compresses the leg veins. Upon arising, she had leg cramps and after walking for 8−10 minutes, she collapsed and died. The findings at autopsy were obesity, mild leg varicosities, an anti-mortem clot in the tibial veins, and short clots of small caliber ledged in the first subdivisions of the pulmonary artery. Simpson recommended that obese or elderly persons be provided with bunks so they could be supine rather than sitting: anticoagulant therapy was not available in London at that time.

During World War II, heparin could be obtained in Sweden from the laboratory of Doctor Jorpes at the Karolinska Institute. H. Zilliacus reported the results of five years' treatment of thrombosis at a series of Swedish hospitals between 1940 and 1945 [113]. Thromboembolism occurred in 1158 of 256,282 (0.45%) patients, with a mortality rate of 14%. However, during the interval from 1941 to 1944, mortality declined from 8.2% to 5.9%, corresponding to an increase in the use of anticoagulant therapy. Data from a single institution (Mariestad Hospital, Sweden) comparing mortality before and after heparin (1929−38 vs 1940−45) showed a 91% decline in mortality and a decrease in fatal embolism from 47 to 3 cases. Zilliacus concluded that specific treatment with heparin limited thrombosis to the lower leg, decreased post-thrombotic sequelae, and helped prevent fatal and non-fatal pulmonary embolism.

In 1954, Marks et al. described the anticoagulant treatment of patients with thrombotic disease admitted to a 500-bed hospital over a 5-year interval [114]. Heparin was given intramuscularly in an initial dose of 150 mg and a coumarin (oral vitamin K antagonist) was given along with the first dose of heparin and continued daily. Subsequent doses of heparin, 50 mg, were given every 4 hours for a total of 4 doses, and hyaluronidase was included with the heparin to decrease pain, local bruising, and improve drug absorption. Heparin monitoring was not considered necessary with the doses of heparin employed, but if monitoring was desired, the therapeutic range of the whole blood clotting time was considered to be 15−25 minutes. This anticoagulation regimen was given to 1135 patients with venous thromboembolism; when compared to historical controls, pain duration declined from 35 days to 3−4 days and confinement in bed from 40 days to 5−7 days. Only 3 patients died of pulmonary embolism if the diagnosis of venous thrombosis was suspected and anticoagulant therapy initiated, compared to 45 fatal emboli in those whose thrombi had not been recognized ante-mortem. The authors also noted that delay in instituting heparin therapy was more likely to result in post-thrombotic sequelae. Reviewing this report, it is curious that the heparin was given intramuscularly rather than intravenously or subcutaneously, which would have avoided the risk of hematoma formation and need for hyaluronidase. In addition, the heparin was discontinued prematurely, before the concentrations of the vitamin-K dependent clotting factors (II, IX, and X) would have decreased sufficiently to prevent thrombus formation. Nevertheless, this report presaged the modern treatment of venous thrombo-embolism and suggested the possibility of treating selected patients at home rather than in hospital.

These and other early studies of heparin's effectiveness were mostly comprised of case-series, but in 1960, Barritt and Jordan [115] published the results of a controlled clinical trial of the use of anticoagulants in 35 patients with pulmonary emboli. Patients were randomized using a blinded allocation system to heparin, 10,000 U intravenously every 6 hours for 6 doses plus a coumarin drug, or no treatment. No monitoring was performed. The data were analyzed one year after the start of the trial, and deaths from pulmonary embolism were increased in the 19 untreated patients compared with the 16 treated (26% vs 0%; P = 0.036). Recurrent pulmonary embolism was also more common in the untreated patients (5 vs 0). The next 38 patients recruited to the trial were all given anticoagulants and no fatalities were recorded. This was a landmark study, showing that giving substantial doses of heparin was effective in preventing recurrent pulmonary emboli. It also confirmed that short courses of heparin are associated with a low risk of serious hemorrhage and might not require monitoring. This protocol was slightly modified by J.G. Sharnoff, who advocated giving 10,000 U the night before surgery [116]. He reported only 2 pulmonary embolism deaths in 1380 major surgical procedures [117].

Surgeons continued to be interested in preventing venous thromboembolism after operations; Vijay Kakkar (1937−2016) made major contributions in this area. He received his medical education in India, continued his training at King's College Hospital in the United Kingdom, and became a fellow of the Royal College of Surgeons in 1964 [118]. His interest in the prophylaxis of venous thrombosis arose after a 3-year-old he had operated upon died of multiple pulmonary emboli. In 1972, Kakkar and associates reported that small doses of heparin (5000 U) could be given subcutaneously prior to and following surgery without incurring increased intraoperative bleeding or postoperative thrombosis [119]. In fact, pulmonary emboli were recognized in only one of 222 patients receiving heparin prophylaxis. This study led to a larger, prospective, clinical trial that randomized 4121 patients undergoing major surgical procedures [120]. Acute, massive pulmonary emboli occurred significantly more often in patients not receiving prophylactic heparin (0.8% vs 0.1%, P < 0.005). There was also a significant decrease in the percentage of patients with leg vein thrombi detected by venography (1.5% vs 0.5%) or by ^{125}I-fibrinogen leg scanning (24.6% vs 7.7%). Analysis of postoperative bleeding showed no statistically significant differences in hemoglobin decline or need for blood transfusions, although more heparin-exposed patients developed wound hematomas ($P < 0.01$). The studies of Kakkar demonstrated that low-dose heparin was safe and effective in preventing thrombosis in surgical patients, and this therapy was soon extended to medical patients as well.

Kakkar and colleagues [121] contributed one of the earliest reports describing the safety and effectiveness of LMWH for the prevention of postoperative thrombosis. They administered a 6,000 kDa heparin fraction or unfractionated heparin to 150 patients undergoing major abdominal operations and reported similar efficacy in prevention venous thromboembolism, and no differences in bleeding. They concluded that a single daily subcutaneous injection of LMWH achieved satisfactory thromboprophylaxis in surgical patients. Subsequently, other trials showed that LMWH was also effective for the treatment of deep vein thrombosis, but excessive bleeding occurred if doses exceeded 1.6 anti-Xa U/mL [122,123].

Kakkar established a Thrombosis Research Institute to further investigate the causes, prevention, and treatment of thrombotic disease. His contributions to improving surgical outcomes were recognized by a Hunterian professorship from the Royal College of Surgeons, a lifetime achievement award from the International Union of Angiology, and, in 2010, the Order of the British Empire.

Fondaparinux and idraparinux

The clinical development of fondaparinux began with six venous thrombosis prevention trials that recruited nearly 9000 patients undergoing hip replacement, hip fracture, and knee replacement. In comparison with enoxaparin, fondaparinux showed equal or superior efficacy with no increase in clinically relevant bleeding [124]. It was FDA-approved for these procedures in 2001; current approvals include thromboembolism prophylaxis in abdominal surgery and the treatment of patients with venous thrombosis and pulmonary embolism.

Idraparinux was compared with standard therapy in 2904 patients with deep vein thrombosis and 2215 with pulmonary embolism [125]. Although equivalent to standard therapy for the former, it was associated with a higher risk for pulmonary embolus recurrence. Idrabiotaparinux was examined in a double-blind, double-dummy, non-inferiority trial versus enoxaparin and warfarin in 3202 patients with acute symptomatic pulmonary embolism [126]. No significant differences were observed in the incidences of recurrent thromboembolism or clinically relevant bleeding. This unique agent was compared with vitamin K antagonists in patients with atrial fibrillation and preliminarily showed similar effectiveness in preventing stroke and systemic embolism, but its development was terminated by the manufacturer for strategic/commercial reasons [127].

Routes of administration

Parenteral administration was not the only route described for achieving anticoagulation. Intrapulmonary instillation of heparin in doses of 1300 U/kg showed rapid clearance from the lung and possibly absorption by the endothelium, with gradual release into the circulation [128]. An anticoagulant effect was observed for up to 14 days. Only a small amount of heparin is detected in plasma after oral administration of customary doses of unmodified drug [129], but if 1000 U/kg is ingested, anti-Xa activity increases within 15 minutes and remains elevated for up to 48 hours [130]. The heparin can be recovered from the urine for as long as 120 hours. Adding sodium N-(8-(2-hydroxybenzoyl) amino)caprylate (SNAC) to heparin facilitates the absorption of therapeutic quantities [131]. A phase II study in patients undergoing total hip replacement showed that thromboprophylaxis with SNAC heparin was as safe and effective as injecting heparin subcutaneously [132]. Heparin is also absorbed when given by the sublingual route, but intramuscular injections are avoided because of the danger of hematoma formation.

As early as 1963, surfaces rinsed with heparin were found to be resistant to clot formation [133], and in subsequent studies, were shown to bind antithrombin and have reduced platelet surface reactivity [134]. Heparin-bonded surfaces have been used for cardiopulmonary bypass equipment and extracorporeal membrane oxygenators [135], and heparin-bonded expanded polytetrafluorethylene grafts are employed for hemodialysis access [136]. A small study reported that heparin-bonded femoral venous catheters were associated with fewer thrombotic complications in children [137], but a Cochrane Review could not confirm that they prolonged the patency of central venous catheters in pediatric patients [138]. Nevertheless, the usefulness of incorporating heparin into catheters, grafts, and other devices continues to be explored.

Monitoring heparin

A reproducible and accurate assay that reflects the potential for bleeding and thrombosis is required to assess the effect of heparin on thrombi and to avoid bleeding. Early workers recorded the time needed for clots to form in blood from cats, dogs, and sheep, as well as man. The whole blood clotting time of Lee and White [139] became widely used, but clot formation in blood from heparinized patients could take as long as 30−60 minutes [140]. Furthermore, it was difficult to judge the moment when a firm clot formed in samples with a prolonged clotting time. The Lee-White clotting time was eventually replaced by a more reproducible and rapid test, the activated partial thromboplastin time (aPTT). This test was originally conceived in 1953 by researchers at the University of North Carolina; Gilbert White has described the history of its development [141,142]. The partial thromboplastin favored by early investigators was a chloroform extract of human brain thromboplastin (tissue factor), and the activator was kaolin [143,144]. To monitor heparin therapy, especially when high concentrations were anticipated, it was recommended that the test plasma be diluted with normal plasma so that the clotting times would fall within an accurately measurable range [145]. The aPTT test has been automated and is used worldwide.

Estes [146] reported that the whole-blood aPTT was a precise and informative assay, but the majority of other workers preferred the platelet-poor plasma aPTT because it was more convenient for routine clinical use. To determine the range of plasma values that coincided with therapeutic heparin levels, Basu et al. [147] performed the aPTT in 243 patients receiving intravenous heparin for the treatment of venous thromboembolism. Doses adjusted to maintain the aPTT between 1.5 and 2.5 times control values prevented recurrent thrombosis, although bleeding occurred in 8% of patients. In another study, Hull et al. [148] reported that patients with aPTT values below the lower limit of the prescribed range (1.5 times the control value) were more likely to have recurrent venous thromboembolism (24.5% vs 1.6%, $P < 0.001$). In addition, patients who failed to achieve a therapeutic aPTT within 24 hours of the initiation of heparin dosing had an increased frequency of recurrent thromboses [149].

Considerable variation in aPTT values among patients receiving the same dose of heparin might be due to differences in the extent of protein binding. One of these proteins, vitronectin, is increased in patients with sepsis [150] and might account for their relative heparin resistance. In addition, the sensitivity of the aPTT to heparin varies with the commercial thromboplastin, contact activation time, and other analytic variables [151]. Lack of predictability of heparin responsiveness resulted in considerable experimentation with heparin dosing. Salzman et al. [152] showed that giving heparin intravenously by continuous infusion, rather than intermittent injection, was effective at preventing thromboembolism and provoked less bleeding. Raschke et al. [153] published a weight-based nomogram for the continuous infusion of heparin (80 U/kg bolus, 18 U/kg per hour) that achieved aPTT values within the therapeutic range more rapidly than using a standard heparin nomogram. Others recommended subcutaneous calcium heparin, either adjusted after an initial dose of 15,000 U to prolong the aPTT to 50−70 s [154], or given in a weight-based regimen of 333 U/kg initially followed by 250 U/kg every 12 hours [155]. Although these measures improved patient outcomes, variability in the sensitivity of laboratory assays to heparin prompted Brill-Edwards et al. [156]. to recommend that each institution establish a therapeutic range for their aPTT using protamine titration, with heparin levels of 0.2−0.4 U/mL as the reference standard.

While the aPTT became the favored assay for monitoring heparin therapy in the clinic, surgeons wishing to prevent bleeding in extravascular circuits and devices in the operating room needed a method of assessing hemostasis that provided immediate results. The accelerated clotting time (ACT) was a point-of-care test introduced by Paul G. Hattersley in 1966 [157]. As originally described, a timer was started when blood entered a silica-containing (diatomite) tube. After thorough mixing, the tube was incubated for 1 minute at 37 °C and then tilted at 5 s intervals until a clot appeared. When used to test blood containing heparin, the ACT was found to be more reproducible, faster, and convenient than the Lee-White Clotting Time. Hattersley used the ACT to monitor intravenous heparin therapy, and reported a satisfactory outcome in all but two of 134 patients if the infusion rate was adjusted to maintain the ACT at 150−190 s [158]. In 1974, Hill et al. [159]. described the application of the ACT to monitor heparin therapy in patients whose blood was being perfused through a membrane oxygenator. They noted that using the ACT to adjust heparin dosage provided more accurate, reliable, and quicker results than the whole blood clotting time. Other investigators observed that the ACT could be employed to guide heparin neutralization with protamine in patients having cardiac operations with cardiopulmonary bypass [160]. Heparin was generally administered in relatively large amounts to prevent clotting in the extracorporeal circuit, and at the termination of bypass, the ACT was performed to determine the appropriate dose of protamine required for resumption of normal hemostasis. The ACT continues to be used for monitoring heparin therapy during cardiac catheterization, coronary bypass/valve replacement, renal hemodialysis, extracorporeal membrane oxygenation, and other procedures, and the test is now fully automated.

Another method available for assessing the anticoagulant effect of heparin was the thrombin clotting time. As originally described by Blomback et al. in 1959, the platelet-poor plasma to be tested is incubated with diluted bovine plasma, and bovine thrombin added [161]. A looped platinum wire is moved up and down in the plasma, and the time when the first fibrin threads appear is noted. With this method, preliminary testing was often necessary to determine the thrombin concentration or dilution of test plasma needed for clotting times within an acceptable range. Later workers simplified the procedure by adding 0.1 mL of bovine thrombin (3 N.I.H. units per mL) to the plasma of heparin-treated patients [162]. They observed that two hours after the subcutaneous administration of 5000 U of heparin to volunteers, the thrombin clotting time increased from a baseline of 20−22 s to >180 s, and returned to initial levels in 6 hours. Although the test was sensitive to low levels of heparin in healthy subjects, it was less accurate in very ill patients because the highly diluted thrombin was affected by potentially high levels of fibrinogen, fibrinogen degradation products, and inhibitors other than heparin in patient plasma, resulting in loss of specificity of the assay [163]. On the other hand, the thrombin clotting time was too sensitive to monitor heparin therapy in patients receiving large amounts of heparin for the treatment of thrombosis; values in these patients were mostly >180 s. Because of the limitations of the thrombin time test, the Wessler laboratory developed an immunoassay for heparin using rabbit antibodies to a methylated bovine serum albumin-heparin precipitate [164]. Plasma heparin concentrations as low as 25 ng/mL could be measured and there was no cross-reactivity with other glycosaminoglycans, but this assay is infrequently used today.

A major advance in the measurement of anticoagulant activity was the anti-Xa assay developed by Yin and coworkers in 1973 [59]. Their approach was based on the ability of heparin to potentiate the inhibition of FXa by antithrombin. Patient plasma was incubated with FXa, clotted with a calcium/phospholipid mixture, and the residual FXa measured. Standard curves were prepared using plasma having known concentrations of heparin. The assay was sensitive to a heparin concentration of as little as 0.01 U/mL. A modification of this assay, the Heptest®, incubates the plasma sample with FXa, followed by a reagent containing calcium chloride and brain cephalin in bovine plasma containing FV and fibrinogen. The clotting time is converted to units of heparin/mL using a standard calibration curve.

We used the anti-Xa assay to monitor heparin therapy in patients receiving aprotinin to reduce blood loss during coronary artery bypass surgery [165]. The ACT could not be used to monitor heparin during bypass because aprotinin contributed to the prolongation of the clotting times, but it did not affect the anti-Xa assay. Blood samples were obtained at 30 minute intervals during bypass, and heparin was dosed based on the anti-Xa assay results; the mean value was 3.6 ± 0.7 U/mL with or without aprotinin. Surgical bleeding was decreased and there was no increase in thrombotic events. Although the further development of aprotinin was curtailed for other reasons, the versatility of the anti-Xa assay was demonstrated. Currently, the method is available in laboratories around the world, with many using commercial kits containing chromogenic or fluorogenic FXa substrates, and the assay has been extremely valuable in drug development. The anti-Xa assay is also used to monitor LMWHs and fondaparinux on the infrequent occasions when clinical assessment of the anticoagulant activity of these agents is indicated.

Heparin analogs

Investigators are continually searching to discover anticoagulants that possess the antithrombotic actions of heparin but do not provoke bleeding. As early as 1977, Thomas et al. [166] described a small glycosaminoglycan polysulfate of bovine origin that potentiated antithrombin but did not alter the activated clotting time. However, after subcutaneous injection, the drug released anti-Xa clotting activity, lipoprotein lipase activity, and platelet factor 4 antigen as well as prolonged the prothrombin time [167]. The *in vivo* effects on coagulation suggest that the analog probably released tissue factor pathway inhibitor. The last publications regarding this agent appeared in 1990, coincident with the general interest and development of LMWHs.

Synthetic heparin mimetics have also attracted much interest; Org 36764, a glycoconjugate [168], and SR 123781A, a hexadecasaccharide with antithrombin and thrombin binding domains [169], were described in 2000–01 but their development was apparently deferred when idraparinux and idrabiotaparinux entered clinical trials. A more recent entrant, EP217609, is a dual action antithrombotic (anti-Xa and anti-IIa) that combines a fondaparinux analog with a thrombin inhibitor (NAPAP) [170]. It has high bioavailability and a long plasma half-life, and was well-tolerated with few adverse events in a phase 1 trial [171]. EP217609 has a biotin tag which enables its rapid and irreversible neutralization by avidin [172] and is being further evaluated in a clinical trial of patients undergoing cardiac surgery. In summary, the overall goal of researchers is to develop anticoagulants whose potency can be adjusted to meet the requirements of the patient, whether for prophylaxis or treatment, and whose activity can be rapidly and safely reversed if necessary.

References

[1] McLean J. The thromboplastic action of cephalin. Am J Physiol 1916;41:250–7.

[2] Howell WH, Holt E. Two new factors in blood coagulation-heparin and pro-antitthrombin. Am J Physiol 1918;47:328–41.

[3] Morawitz P. The chemistry of blood coagulaton. Ergebn Physiol 1905;4:307–411. Reprinted in Hartmann RC, Guenther PF (trans), Springfield, Ill, Charles C Thomas, 1958, 194 pp.

[4] Howell WH. The purification of heparin and its presence in blood. Am J Physiol 1925;71:553–62.

[5] Best CH. Preparation of heparin and its use in the first clinical cases. Circulation 1959;19:79–86.

[6] Charles AF, Scott IV DA. Observations on the chemistry of heparin. Biochem J 1936;30:1927–33.

[7] Shampo MA, Kyle RA. J. Erik Jorpes-pioneer in the identification and clinical applications of heparin. Mayo Clin Proc 1997;72:1056.

[8] Barrowclife TW. History of heparin. In: Lever R, Mulloy B, Page CP, editors. Heparin-a century of progress. Berlin Heidelberg: Springer-Verlag; 2012. p. 3–22.

[9] Jorpes E. The chemistry of heparin. Biochem J 1935;29:1817–30.

[10] Jorpes E, Bergstrom SVII. On the relationship between the sulphur content and the anticoagulant activity of heparin preparations. Biochem J 1939;33:47–52.

[11] Jaques LB. The heparins of various mammalian species and their relative anti-coagulant potency. Science 1940;92:488–9.

[12] Jaques LB. Heparin: an old drug with a new paradigm. Science 1979;206:528–33.

[13] Cifonelli JA, Dorfman A. The uronic acid of heparin. Biochem Biophys Res Commun 1962;7:41–5.

[14] Jaques LB. The mast cells in the light of new knowledge of heparin and sulfated mucopolysaccharides. Gen Pharmac 1975;6:235–45.

[15] Metcalfe DD, Lewis RA, Silbert JE, Rosenberg RD, Wasserman SI, Austen KF. Isolation and characterization of heparin from human lung. J Clin Invest 1979;64:1537–43.

[16] Carlsson P, Kjellen L. Heparin biosynthesis. In: Lever R, Mulloy B, Page CP, editors. Heparin-a century of progress. Berlin Heidelberg: Springer-Verlag; 2012. p. 23–41.

[17] Forsberg E, Pejler G, Ringvall M, Lunderius C, Tomasini-Johansson B, Kusche-Gullberg M, et al. Abnormal mast cells in mice deficient in a heparin-synthesizing enzyme. Nature 1999;400:773–6.

[18] Dreyfuss JL, Regatieri CV, Jarrouge TR, et al. Heparan sulfate proteoglycans: structure, protein interactions and cell signaling. An Acad Bras Cienc 2009;81:409–29.

[19] Zaferani A, Talsma D, Richter MKS, Daha MR, Navis GJ, Seelen MA, et al. Heparin/heparan sulphate interactions with complement—a possible target for reduction of renal function loss? Nephrol Dial Transplant 2014;29:515–22.

[20] Lindahl U, Kjellén L. Pathophysiology of heparan sulphate: many diseases, few drugs. J Intern Med 2013;273:555–71.

[21] Gallagher JT, Walker A. Molecular distinctions between heparan-sulfate and heparin. Analysis of sulfation patterns indicates that heparan-sulfate and heparin are separate families of N-sulfated polysaccharides. Biochem J 1985;230:665–74.

[22] Li L, Ly M, Linhardt RJ. Proteoglycan sequence. Mol Biosyst 2012;8:1613–25.

[23] Olczyk P, Mencner L, Komosinska-Vassev K. Diverse roles of heparan sulfate and heparin in wound repair. Biomed Res Int 2015;2015:549417.

[24] Harenberg J. Pharmacology of low molecular weight heparins. Semin Thromb Haemost 1990;16(Suppl):12–18.

[25] Petitou M, van Boeckel CA. A synthetic antithrombin III binding pentasaccharide is now a drug! What comes next? Angew Chem Int Ed Engl 2004;43:3118–33.

[26] Herbert JM, Herault JP, Bernat A, van Amsterdam RGM, Lormeau JC, Petitou M, et al. Biochemical and pharmacological properties of SANORG 34006, a potent and long-acting synthetic pentasaccharide. Blood 1998;91:4197–205.

[27] Savi P, Herault JP, Duchaussoy P, Millet L, Schaeffer P, Petitou M, et al. Reversible biotinylated oligosaccharides: a new approach for a better management of anticoagulant therapy. J Thromb Haemost 2008;6:1697–706.

[28] The Equinox Investigators. Efficacy and safety of once weekly subcutaneous idrabiotaparinux in the treatment of patients with symptomatic deep venous thrombosis. J Thromb Haemost 2011;9:92–9.

[29] Turpie AG, Robinson JG, Doyle DJ, Mulji AS, Mishkel GJ, Sealey BJ, et al. Comparison of high-dose with low-dose subcutaneous heparin to prevent left ventricular mural thrombosis in patients with acute transmural anterior myocardial infarction. N Engl J Med 1989;320:352–7.

[30] Thomas DP, Sagar S, Stamatakis JD, Maffei FHA, Erdi A, Kakkar VV. Plasma heparin levels after administration of calcium and sodium salts of heparin. Thromb Res 1976;9:241–8.

[31] Kroon C, de Boer A, Kroon JM, Schoenmaker HC, vd Meer FJM, Cohen AF. Influence of skinfold thickness on heparin absorption. Lancet 1991;337:945–6.

[32] Young E, Prins M, Levine MN, Hirsh J. Heparin binding to plasma proteins, an important mechanism for heparin resistance. Thromb Haemost 1992;67:639–43.

[33] Manson L, Weitz JI, Podor TJ, Hirsh J, Young E. The variable anticoagulant response to unfractionated heparin in vivo reflects binding to plasma proteins rather than clearance. J Lab Clin Med 1997;130:649−55.

[34] Oie Cl, Olsen R, Smedsrod B, Hansen JB. Liver sinusoidal endothelial cells are the principal site for elimination of unfractionated heparin from the circulation. Am J Physiol Gastrointest Liver Physiol 2008;294:G520−8.

[35] Harris EN, Baggenstoss BA, Weigel PH. Rat and human HARE/stabilin-2 are clearance receptors for high- and low-molecular-weight heparins. Am J Physiol Gastrointest Liver Physiol 2009;296:G1191−9.

[36] Pandey MS, Harris EN, Weigel PH. HARE-mediated endocytosis of hyaluronan and heparin is targeted by different subsets of three endocytic motifs. Int J Cell Biol 2015;2015:524707.

[37] Hirsh J, Warkentin TE, Shaughnessy SG, Anand SS, Halperin JL, Raschke R, et al. Heparin and low-molecular weight heparin. Mechanisms of action, pharmacokinetics, dosing, monitoring, efficacy, and safety. Chest 2001;119:64S−94S.

[38] Olsson P, Lagergren H, Ek S. The elimination from plasma of intravenous heparin: an experimental study on dogs and humans. Acta Med Scand 1963;173:619−30.

[39] De Swart CAM, Nijmeyer B, Roelofs JMM, Sixma JJ. Kinetics of intravenously administered heparin in normal humans. Blood 1982;60:1251−8.

[40] Young E, Wells P, Holloway S, Weitz J, Hirsh J. Ex-vivo and in-vitro evidence that low molecular weight heparins exhibit less binding to plasma proteins than unfractionated heparin. Thromb Haemost 1994;71:300−4.

[41] Boneu B, Caranobe C, Sie P. Pharmacokinetics of heparin and low molecular weight heparin. Baillieres Clin Haematol 1990;3:531−44.

[42] Collignon F, Frydman A, Caplain H, Ozoux ML, Le Roux Y, Bouthier J, et al. Comparison of the pharmacokinetic profiles of three low molecular mass heparins-dalteparin, enoxaparin and nadroparin-administered subcutaneously in healthy volunteers (doses for prevention of thromboembolism). Thromb Haemost 1995;73:630−40.

[43] McAllister BM, Demis DJ. Heparin metabolism: isolation and characterization of uroheparin. Nature 1966;212(5059):293−4.

[44] Kikkert WJ, van Brussel PM, Damman P, Claessen BI, van Straalen JP, Vis MM, et al. Influence of chronic kidney disease on anticoagulation levels and bleeding after primary percutaneous coronary intervention in patients treated with unfractionated heparin. J Thromb Thrombolysis 2016;41:441−51.

[45] Mahé I, Aghassarian M, Drouet L, Bal Dit-Sollier C, Lacut K, Heilmann JJ, et al. Tinzaparin and enoxaparin given at prophylactic dose for eight days in medical elderly patients with impaired renal function: a comparative pharmacokinetic study. Thromb Haemost 2007;97:581−6.

[46] Lim W, Dentali F, Eikelboom JW, Crowther MA. Meta-analysis: low-molecular-weight heparin and bleeding in patients with severe renal insufficiency. Ann Intern Med 2006;144:673−84.

[47] Siguret V, Gouin-Thibault I, Pautas E, Leizorovicz A. No accumulation of the peak anti-factor Xa activity of tinzaparin in elderly patients with moderate-to-severe renal impairment: the IRIS substudy. J Thromb Haemost 2011;9:1966−72.

[48] Leizorovicz A, Siguret V, Mottier D, Innohep® in Renal Insufficiency Study Steering Committee, Leizorovicz A, Siguret V, et al. Safety profile of tinzaparin versus subcutaneous unfractionated heparin in elderly patients with impaired renal function treated for

acute deep vein thrombosis: the Innohep® in Renal Insufficiency Study (IRIS). Thromb Res 2011;128:27−34.

[49] Arixtra package insert.

[50] Brinkhous KM, Smith HP, Warner ED, Seegers WH. The inhibition of blood clotting: an unidentified substance which acts in conjunction with heparin to prevent the conversion of prothrombin into thrombin. Am J Physiol 1939;125:683−7.

[51] Waugh DF, Fitzgerald MA. Quantitative aspects of antithrombin and heparin in plasma. Am J Physiol 1956;184:627−39.

[52] Abildgaard U. Inhibition of the thrombin-fibrinogen reaction by heparin and purified cofactor. Scand J Haematol 1968;5:440−53.

[53] Ellis V, Scully MF, Kakkar VV. The acceleration of the inhibition of platelet prothrombinase complex by heparin. Biochem J 1986;233:161−5.

[54] Rosenberg RD, Damus PS. The purification and mechanism of action of human antithrombin-heparin cofactor. J Biol Chem 1973;248:6490−505.

[55] Rosenberg RD, Lam L. Correlation between structure and function of heparin. Proc Natl Acad Sci USA 1979;76:1218.

[56] Choay J, Petitou M, Lormeau JC, Sinay P, Casu B, Gatti G. Structure-activity relationship in heparin: a synthetic pentasaccharide with high affinity for antithrombin III and eliciting high anti-factor Xa activity. Biochem Biophys Res Commun 1983;116:492−9.

[57] Lindahl U, Bäckström G, Thunberg L, Leder IG. Evidence for a 3-O-sulfated D-glucosamine residue in the antithrombin-binding sequence of heparin. Proc Natl Acad Sci USA 1980;77:6551−5.

[58] Denson KWE, Bonnar J. The measurement of heparin. A method based on the potentiation of anti-factor Xa. Thromb Diath Haemorrh 1973;30:471−9.

[59] Yin ET, Wessler S, Butler JV. Plasma heparin: a unique, practical, submicrogram-sensitive assay. J Lab Clin Med 1973;81:298−310.

[60] Andersson LO, Barrowcliffe TW, Holmer E, Johnson EA, Söderström G. Molecular weight dependency of the heparin potentiated inhibition of thrombin and activated factor X. Effect of heparin neutralization in plasma. Thromb Res 1979;15:531−41.

[61] Lane DA, Denton J, Flynn AM, Thunberg L, Lindahl U. Anticoagulant activities of heparin oligosaccharides and their neutralization by platelet factor 4. Biochem J 1984;218:725−32.

[62] Johnson EA, Kirkwood TB, Stirling Y, Perez-Requejo JL, Ingram GI, Bangham DR, et al. Four heparin preparations: anti-Xa potentiating effect of heparin after subcutaneous injection. Thromb Haemost 1976;35:586−91.

[63] Lormeau JC, Herault JP. The effect of the synthetic pentasaccharide SR 90107/ORG 31540 on thrombin generation ex vivo is uniquely due to ATIII-mediated neutralization of factor Xa. Thromb Haemost 1995;74:1474−7.

[64] Gray E, Hogwood J, Mulloy B. The anticoagulant and antithrombotic mechanisms of heparin. In: Lever R, Mulloy B, Page CP, editors. Heparin-a century of progress. Berlin Heidelberg: Springer-Verlag; 2012. p. 43−61.

[65] Gettins PG, Olson ST. Inhibitory serpins. New insights into their folding, polymerization, regulation and clearance. Biochem J 2016;473:2273−93.

[66] Carlström AS, Liedén K, Björk I. Decreased binding of heparin to antithrombin following the interaction between antithrombin and thrombin. Thromb Res 1977;11:785−97.

[67] Al Dieri R, Wagenvoord R, van Dedem GW, Beguin S, Hemker HC. The inhibition of blood coagulation by heparins of different molecular weight is caused by a common functional motif-the C-domain. J Thromb Haemost 2003;1:907−14.

[68] Wagenvoord R, Al Dieri R, van Dedem G, Beguin S, Hemker HC. Linear diffusion of thrombin and factor Xa along the heparin molecule explains the effects of extended heparin chain lengths. Thromb Res 2008;122:237−45.

[69] Hemker HC. A century of heparin: past, present and future. J Thromb Haemost 2016;14:2329−38.

[70] Hemker HC, Giesen P, Al Dieri R, Regnault V, de Smed E, Wagenvoord R, et al. The calibrated automated thrombogram (CAT): a universal routine test for hyper− and hyprocoagulability. Pathophysiol Haemost Thromb 2002;32:249−53.

[71] Gerotziafas GT, Petropoulou AD, Verdy E, Samama MM, Elalamy I. Effect of the antifactor Xa and anti-factor IIa activities of low-molecular-weight heparins upon the phases of thrombin generation. J Thromb Haemost 2007;5:955−62.

[72] Krulder JWM, Strebus AF, Meinders AE, Briet E. Anticoagulant effect of unfractionated heparin in antithrombin-depleted plasma in vitro. Haemostasis 1996;26:85−9.

[73] Sandset PM, Abildgaard U, Larsen ML. Heparin induces release of extrinsic coagulation pathway inhibitor (EPI). Thromb Res 1988;50:803−13.

[74] Lindahl AK, Abildgaard U, Stokke G. Release of extrinsic pathway inhibitor after heparin injection: increased response in cancer patients. Thromb Res 1990;59:651−6.

[75] Nakamura R, Umemura K, Hashimmoto H, Urano T. Less pronounced enhancement of thrombin-dependent inactivation of plasminogen activator inhibitor type 1 by low molecular weight heparin. Thromb Haemost 2006;95:637−42.

[76] Fry ETA, Sobel BE. Lack of interference by heparin with thrombolysis or binding of tissue-type plasminogen activator to thrombin. Blood 1988;71:1347−52.

[77] Xu X, Takano R, Nagai Y, Yanangida T, Kamei K, Kato H, et al. Effect of heparin chain length on the interaction with tissue factor pathway inhibitor (TFPI). Int J Bio Macromol 2002;30:151−60.

[78] Fernandez JA, Petaja J, Griffin JH. Dermatan sulfate and LMW heparin enhance the anticoagulant action of activated protein C. Thromb Haemost 1999;82:1462−8.

[79] Leung L, Saigo K, Grant D. Heparin binds to human monocytes and modulates their procoagulant activities and secretory phenotypes. Effect of histidine-rich glycoprotein. Blood 1989;73:177−84.

[80] Rosenberg RD. Heparin action. Circulation 1974;49:603−5.

[81] Von dem Borne PA, Meijers JCM, Bouma BN. Effect of heparin on the activation of factor XI by fibrin-bound thrombin. Thromb Haemost 1996;76:347−53.

[82] Eika C. Inhibition of thrombin-induced aggregation of human platelets by heparin. Scand J Haemat 1971;8:216−22.

[83] Eika C. On the mechanism of platelet aggregation induced by heparin, protamine and polybrene. Scand J Haemat 1972;9:248−57.

[84] Gao C, Boylan B, Fang J, Wilcox DA, Newman DK, Newman PJ. Heparin promotes platelet responsiveness by potentiating $\alpha IIb\beta 3$-mediated outside-in-signaling. Blood 2011;117:4946−52.

[85] Heiden D, Mielke Jr CH, Rodvien R. Impairment by heparin of primary hemostasis and platelet ^{14}C-5-hydroxytryptamine release. Br J Haematol 1977;36:427−36.

[86] Shanberge JN, Kambayashi J, Nakagawa M. The interaction of platelets with a tritium-labelled heparin. Thromb Res 1976;595−609.

[87] Messmore HL, Griffin B, Koza M, Seghatchian J, Fareed J, Coyne E. Interaction of heparinoids with platelets: comparison with heparin and low molecular weight heparins. Semin Thromb Hemost 1991;17(Suppl 1):57−9.

[88] Ferrer-Lopez P, Renesto P, Prevost M-C, Gounon P, Chignard M. Heparin inhibits neutrophil-induced platelet activation via cathepsin G. J Lab Clin Med 1992;119:231−9.

[89] Eslin DE, Zhang C, Samuels KJ, Rauova L, Zhai L, Niewiarowski S, et al. Transgenic mice studies demonstrate a role for platelet factor 4 in thrombosis: dissociation between anticoagulant and antithrombotic effect of heparin. Blood 2004;104:3173−80.

[90] Sobel M, Soler DF, Kermode JC, Harrris RB. Localization and characterization of a heparin binding domain peptide of human von Willebrand factor. J Biol Chem 1992;267:8857−62.

[91] Kroner PA, Frey AB. Analysis of the structure and function of the von Willebrand factor A1 domain using targeted deletions and alanine-scanning mutagenesis. Biochemistry 1996;35:13460−8.

[92] Perrault C, Ajzenberg N, Legendre P, Rastegar-Lari G, Meyer D, Lopez JA, et al. Modulation by heparin of the interaction of the A1 domain of Von Willebrand factor with glycoprotein Ib. Blood 1999;94:4186−94.

[93] Badimon L, Badimon JJ, Lassilia R, Heras M, Chesebro JH, Fuster V. Thrombin regulation of platelet interaction with damaged vessel wall and isolated collagen type 1 at arterial flow conditions in a porcine model: effects of hirudins, heparin, and calcium chelation. Blood 1991;78:423−34.

[94] Cadroy Y, Harker LA, Hanson SR. Inhibition of platelet-dependent thrombosis by low molecular weight heparin (CY222): comparison with standard heparin. J Lab Clin Med 1989;114:349−57.

[95] Chargaff E, Olson VI KB. Studies on the action of heparin and other anticoagulants. The influence of protamine on the anticoagulant effect in vivo. J Biol Chem 1937;122:153−67.

[96] Bhaskar U, Sterner E, Hickey AM, Onishi A, Zhang F, Dordick JS, et al. Engineering of routes to heparin and related polysaccharides. Appl Microbiol Biotechnol 2012;93:1−16.

[97] Marcum JA, Atha DH, Fritze LM, Nawroth P, Stern D, Rosenberg RD. Cloned bovine aortic endothelial cells synthesize anticoagulantly active heparan sulfate proteolglycan. J Biol Chem 1986;261:7507−17.

[98] Marcum JA, McKenney JB, Galli SJ, Jackman RW, Rosenberg RD. Anticoagulantly active heparin-like molecules from mast cell-deficient mice. Am J Physiol 1986;250:H879−88.

[99] Staprans I, Felts JM. Isolation and characterization of glycosaminoglycans in human plasma. J Clin Invest 1985;76:1984−91.

[100] Barrowcliffe TW, Johnson EA, Eggleton CA, Thomas DP. Anticoagulant activities of lung and mucous heparins. Thromb Res 1978;12:27−36.

[101] Jaques LB. Standardisation of heparin for clinical use. Lancet 1975;i:287.

[102] Gray E. Standardisation of unfractionated and low-molecular-weight heparin. In: Lever R, Mulloy B, Page CP, editors. Heparin-a century of progress. Berlin Heidelberg: Springer-Verlag; 2012. p. 65−76.

[103] Triplett DA. Heparin: biochemistry, therapy, and laboratory monitoring. Ther Drug Monit 1979;1:173−97.

[104] Jaques LB, Kavanagh LW. Standardisation of heparin for clinical use. Lancet 1972;ii:1315.

[105] Interim revision announcement. Pharmacopeial Forum 2009;35:1−7.

[106] Barrowcliffe TW, Curtis AD, Tomlinson STP, Hubbard AR, Johnson EA, Thomas DP. Standardization of low molecular weight heparins: a collaborative study. Thromb Haemost 1985;54:675−9.

[107]   Barrowcliffe TW, Curtis AD, Johnson EA, Thomas DP. An international standard for low molecular weight heparins. Thromb Haemost 1988;60:1—7.

[108]   Gray E, Rigsby P, Behr-Gross ME. Collaborative study to establish the low-molecular—mass heparin for assay—European Pharmacopoeia Biological Reference Preparation. Pharmeuropa Bio 2004;2004:59—76.

[109]   Terao E, Daas A. Collaborative study for the calibration of replacement batches for the heparin low-molecular-mass for assay biological reference preparation. Pharmeur Bio Sci Notes 2016;2015:35—47.

[110]   Jorpes JE. Heparin: a mucopolysaccharide and an active antithrombotic drug. Circulation 1959;19:87—91.

[111]   Murray DWG, Best CH. The use of heparin in thrombosis. Ann Surg 1938;108:163—77.

[112]   Simpson K. Shelter deaths from pulmonary embolism. Lancet 1940;ii:744.

[113]   Zilliacus H. On the specific treatment of thrombosis and pulmonary embolism with anticoagulants, with particular reference to the post-thrombotic sequelae. Acta Med Scand 1946;Suppl 171:51—171.

[114]   Marks J, Truscott BM, Withycombe JFR. Treatment of venous thrombosis with anticoagulants. Lancet 1954;ii:787—91.

[115]   Barritt DW, Jordan SC. Anticoagulant drugs in the treatment of pulmonary embolism. Lancet 1960;i:1309—12.

[116]   Sharnoff JG, Kass HH, Mistica BA. A plan of heparinization of the surgical patient to prevent postoperative thromboembolism. Surg Gynecol Obstet 1962;115:75—9.

[117]   Sharnoff JG. Venous thromboembolism. N Engl J Med 1972;287:1201.

[118]   Anonymous. Vijay Kakkar. Br Med J 2017;356:6852.

[119]   Kakkar VV, Spindler J, Flute PT, Corrigan T, Fossard DP, Crelin RQ, et al. Efficacy of low doses of heparin in prevention of deep-vein thrombosis after major surgery: a double-blind, randomized trial. Lancet 1972;300:101—6.

[120]   Kakkar VV, Corrigan TP, Fossard DP, Sutherland I, Shelton MG, Thirlwall J. Prevention of fatal postoperative pulmonary embolism by low doses of heparin. Lancet 1975;306:45—64.

[121]   Kakkar VV, Djazaeri B, Fok J, Fletcher M, Scully MF, Westwick J. Low-molecular-weight heparin and prevention of postoperative deep vein thrombosis. Br Med J 1982;284(6313):375—9.

[122]   Bratt G, Törnebohm E, Granqvist S, Aberg W, Lockner D. A comparison between low molecular weight heparin (KABI 2165) and standard heparin in the intravenous treatment of deep venous thrombosis. Thromb Haemost 1985;54:813—17.

[123]   Koller M, Schoch U, Buchmann P, Largiader F, von Felten A, Frick PG. Low molecular weight heparin (KABI 2165) as thromboprophylaxis in elective visceral surgery. A randomized, double-blind study versus unfractionated heparin. Thromb Haemost 1986;56:243—6.

[124]   Turpie AGG, Eriksson BI, Bauer KA, Lassen MR. New pentasaccharides for the prophylaxis of venous thromboembolism. Chest 2003;124(Suppl):371S—8S.

[125]   The van Gogh Investigators. Idraparinux versus standard therapy for venous thromboembolic disease. N Engl J Med 2007;357:1094—104.

[126]   Büller HR, Gallus AS, Pillion G, Prins MH, Raskob GE. Cassiopea investigators. Enoxaparin followed by once-weekly idrabiotaparinux versus enoxaparin plus warfarin for patients with acute symptomatic pulmonary embolism: a randomised, double-blind, double-dummy, non-inferiority trial. Lancet 2012;379.123—9.

[127]   Buller HR, Halperin J, Hankey GJ, Pillion G, Prins MH, Raskob GE. Comparison of idrabiotaparinux with vitamin K antagonists for prevention of thromboembolism in patients with atrial fibrillation: the Borealis-atrial fibrillation study. J Thromb Haemost 2014;12:824−30.

[128]   Jaques LB, Mahadoo J, Kavanagh LW. Intrapulmonary heparin: a new procedure for anticoagulant therapy. Lancet 1976;ii:1157−61.

[129]   Jaques LB, Hiebert LM, Wice SM. Evidence from endothelium of gastric absorption of heparin and of dextran sulfates 8000. J Lab Clin Med 1991;117:122−30.

[130]   Hebert LM, Wice SM, Ping T. Increased plasma anti-Xa activity and recovery of heparin from urine suggest absorption of orally administered unfractionated heparin in human subjects. J Lab Clin Med 2005;145:151−5.

[131]   Pineo G, Hull R, Marder V. Oral delivery of heparin: SNAC and related formulations. Best Pract Res Clin Haematol 2004;17:153−60.

[132]   Pineo GF, Hull RD, Marder VJ. Orally active heparin and low-molecular-weight heparin. Curr Opin Pulm Med 2001;7:344−8.

[133]   Gott VL, Whiffen JD, Dutton RC. Heparin bonding on colloidal graphite surfaces. Science 1963;142:1297−8.

[134]   Lindon JN, Salzman EW, Merrill EW, Dincer AK, Labarre D, Bauer KA, et al. Catalytic activity and platelet reactivity of heparin covalently bonded to surfaces. J Lab Clin Med 1985;105:219−26.

[135]   von Segesser LK. Heparin-bonded surfaces in extracorporeal membrane oxygenation for cardiac support. Ann Thorac Surg 1996;61:330−5.

[136]   Olsha O, Goldin I, Shemesh D. Heparin-bonded expanded polytetrafluorethylene grafts in hemodialysis access. J Vasc Assess 2016;17(Suppl 1):S79−84.

[137]   Krafte-Jocobs B, Sivit CJ, Meija R, Pollack MM. Catheter-related thrombosis in critically ill children: comparison of catheters with and without heparin bonding. J Pediatr 1995;126:50−4.

[138]   Shah PS, Shah N. Heparin-bonded catheters for prolonging the patency of central venous catheters in children. Cochrane Database Sys Rev 2014;25:CD005983.

[139]   Lee RI, White PD. A clinical study of the coagulation time of blood. Am J Med Sci 1913;145:495−503.

[140]   Marple CD. The administration of anticoagulants. Calif Med 1950;73:166−70.

[141]   Langdell RD, Wagner RH, Brinkhous KM. Effect of anti-hemophilic factor on one-stage clotting tests: a presumptive test for hemophilia and a simple one-stage anti-hemophilic factor assay procedure. J Lab Clin Med 1953;41:637−47.

[142]   White GC. The partial thromboplastin time: defining an era in coagulation. J Thromb Haemost 2003;1:2267−70.

[143]   Bell WN, Alton HG. A brain extract as a substitute for platelet suspensions in the thromboplastin generation time. Nature 1954;174:880−1.

[144]   Proctor RR, Rapaport SI. The partial thromboplastin time with kaolin. A simple screening test for first stage plasma clotting factor deficiencies. Am J Clin Path 1961;36:212−19.

[145]   Marder VJ. A simple technique for the measurement of plasma heparin concentration during anticoagulant therapy. Thromb Diath Haemorrh 1970;24:230−9.

[146]   Estes JW. Kinetics of the anticoagulant effect of heparin. JAMA 1970;212:1492−5.

[147]   Basu D, Gallus A, Hirsh J, Cade J. A prospective study of the value of monitoring heparin treatment with the activated partial thromboplastin time. N Engl J Med 1972;287:324−7.

[148] Hull RD, Raskob GE, Hirsh J, Jay RM, Leclerc JR, Geerts WH, et al. Continuous intravenous heparin compared with intermittent subcutaneous heparin in the initial treatment of proximal-vein thrombosis. N Engl J Med 1986;315:1109—14.

[149] Hull RD, Raskob GE, Brant RF, Pineo GF, Valentine KA. Relation between the time to achieve the lower limit of the aPTT therapeutic range and recurrent venous thromboembolism during heparin treatment for deep vein thrombosis. Arch Intern Med 1997;157:2562—8.

[150] Young E, Podor TJ, Venner T, Hirsh J. Induction of the acute-phase reaction increases heparin-binding proteins in plasma. Arterioscler Thromb Vasc Biol 1997;17:1568—74.

[151] Owen J, Payne E, Carstairs K. Erroneous activated partial thromboplastin time. Ann Intern Med 1978;89:146—7.

[152] Salzman EW, Deykin D, Shapiro RM, Rosenberg R. Management of heparin therapy: controlled prospective trial. N Engl J Med 1975;292:1046—50.

[153] Raschke RA, Reilly BM, Guidry JR, Fontana JR, Srinivas S. The weight-based heparin dosing nomogram compared with a "standard care" nomogram. Ann Intern Med 1993;119:874—81.

[154] Doyle DJ, Turpie AG, Hirsh J, Best C, Kinch D, Levine MN, et al. Adjusted subcutaneous heparin or continuous intravenous heparin in patients with acute deep vein thrombosis. A randomized trial. Ann Intern Med 1987;107:441—5.

[155] Kearon C, Ginsberg JS, Julian JA, Douketis J, Solymoss S, Ockelford P, et al. Fixed-dose heparin (FIDO) investigators. Comparison of fixed-dose weight-adjusted unfractionated heparin and low-molecular-weight heparin for acute treatment of venous thromboembolism. JAMA 2006;296:935—42.

[156] Brill-Edwards P, Ginsberg JS, Johnston M, Hirsh J. Establishing a therapeutic range for heparin therapy. Ann Intern Med 1993;119:104—9.

[157] Hattersley PG. Activated coagulation time of whole blood. JAMA 1966;196:436—40.

[158] Hattersley PG, Mitsuoka JC, King JH. Heparin therapy for thromboembolic disorders. A prospective evaluation of 134 cases monitored by the activated coagulation time. JAMA 1983;250:1413—16.

[159] Hill JD, Dontigny L, de Leval M, Mielke Jr. CH. A simple method of heparin management during prolonged extracorporeal circulation. Ann Thorac Surg 1974;17:129—33.

[160] Mattox KL, Guinn GA, Rubio PA, Beall Jr. AC. Use of the activated coagulation time in intraoperative heparin reversal for cardiopulmonary operations. Ann Thorac Surg 1975;19:634—8.

[161] Blomback B, Blomback M, Olsson P, William-Olsson G, Senning A. Determination of heparin level of the blood. Acta Chir Scand Suppl 1959;245:259—64.

[162] Eika C, Godal HC, Kierulf P. Detection of small amounts of heparin by the thrombin clotting-time. Lancet 1973;2(7773):376.

[163] Wessler S, Yin ET, Flute PT, Kakkar VV. Thrombin-time test. Lancet 1972;2:877—8.

[164] Gitel SN, Medina VM, Wessler S. Antiheparin antibodies: their preparation and use in a heparin immunoassay. J Lab Clin Med 1987;109:672—8.

[165] Green D, Sanders J, Eiken M, Wong CA, Frederiksen J, Joob A, et al. Recombinant aprotinin in coronary artery bypass graft operations. J Thorac Cardiovasc Surg 1995;110:963—70.

[166] Thomas DP, Michalski R, Lane DA, Johnson EA, Kakkar VV. A heparin analogue with specific action on antithrombin III. Lancet 1977;i:120—2.

[167] Kakkar VV, Djazaeri B, Scully M, Weerasinghe K. Synthetic heparin analogue and prothrombin time. Lancet 1981;i:1167.

[168] Vogel GMT, van Amsterdam GM, van Dinther G, Tromp M, Meuleman DG. Pre-clinical pharmacological profile of the novel glycoconjugate Org 36764 with both factor Xa and thrombin (IIa) inhibitory activities. Thromb Haemost 2000;84:611−20.

[169] Herbert JM, Herault JP, Bernat A, Savi P, Schaeffer P, Driquez PA. Duchaussoy Ph, Petitou M. SR123781A, a synthetic heparin mimetic. Thromb Haemost 2001;85:852−60.

[170] Olson ST, Swanson R, Petitou M. Specificity and selectivity profile of EP217609: a new neutralizable dual-action anticoagulant that targets thrombin and factor Xa. Blood 2012;119:2187−95.

[171] Gueret P, Combe S, Krezel C, Fuseau E, van Giersbergen PL, Petitou M, et al. First in man study of EP217609, a new long-acting, neutralisable parenteral antithrombotic with a dual mechanism of action. Eur J Clin Pharmacol 2016;72:1041−50.

[172] Gueret P, Combe S, Krezel C, Fuseau E, van Giersbergen PL, Petitou M, et al. Neutralization of EP217609, a new dual-action FIIa/FXa anticoagulant, by its specific antidote avidin: a phase I study. Eur J Clin Pharmacol 2017;73:15−28.

Chapter 2

Non-anticoagulant heparins

In Chapter 1, it was noted that heparin resides in mast cell granules as well as on the surface of the endothelium. It is unlikely to function as an anticoagulant under ordinary circumstances because only small amounts are found in the circulating blood. However, heparin has many other properties that might be relevant to its physiologic activities in health and disease. In 1993, Lane and Adams wrote an editorial in the New England Journal of Medicine describing a few of the non-anticoagulant functions of heparin [1]. They noted that heparin can affect cell signaling and adhesion by binding to a variety of extracellular matrix proteins, release lipoprotein lipase from the vessel wall, and modify inflammation by interacting with complement proteins. Because of these features, they suggested that heparins might have therapeutic applications beyond the management of thrombosis.

In this Chapter, several of the non-anticoagulant functions of heparin will be reviewed. However, one must be cautious in interpreting the results of the experimental data because heparin preparations are heterogeneous. Depending on the particular tissue and extraction process used by the manufacturer, the final product can vary in the number of saccharide units and extent of sulfation, factors that critically affect its non-anticoagulant as well as anticoagulant properties. Reducing the number of saccharide units to produce low molecular weight heparins (LMWHs) might remove structural elements required for specific non-anticoagulant activities [2]. Decreasing the number of saccharide units from 20 to 4 resulted in a stepwise decline in the percent inhibition of neutrophil elastase release from 88% to 39% [3]. Another study that directly compared the various commercial heparins reported inhibition of heparanase was greatest by unfractionated heparin (95%), moderate by two LMWHs (tinzaparin, 89%; dalteparin, 77%) and least with enoxaparin (58%) [4]. Although these heparins had similar anti-Xa activity (163–188 U/mg), the molecular mass of tinzaparin exceeds that of enoxaparin (6500 vs 4500 g/mol). Furthermore, the non-anticoagulant activities might vary among the LMWHs because each is prepared by a distinct patented process. If LMWH is depolymerized by heparinase I, it is capable of suppressing TNF-α and IL-1β secretion, but if other heparinases (II or III) are used for the depolymerization, this activity is decreased or entirely absent [5]. In addition, the particular characteristic being evaluated depends on

The Heparins. DOI: https://doi.org/10.1016/B978-0-12-818781-4.00002-9

35

whether testing is being performed in a chemically-defined solution, in plasma, or in other fluids, because of the strong tendency of heparins to form complexes with coexistent proteins or other substances. Furthermore, when heparin enters the circulation it is competing with vessel wall heparans and heparanase for binding to various targets. In the following paragraphs, the generally accepted non-anticoagulant functions of heparin will be described.

Release of lipases

In 1943, Hahn reported that post-prandial lipemia in dogs cleared within 1−3 minutes after an intravenous infusion of heparin, but clearing did not occur if lipemic blood was mixed with heparin *in vitro* [6]. The *in vivo* clearing factor had the characteristics of a protein [7] and its clearing activity was attributed to the hydrolysis of triglycerides similar to the action of pancreatic lipase on glyceride ester bonds [8]. These studies identified it as a lipoprotein lipase (LPL), and further studies in human subjects showed that raised levels of LPL persist for as long as 24 hours after a 50 mg subcutaneous dose of heparin [9]. Non-esterified fatty acids are released but there is no decrease in the levels of circulating triglycerides, cholesterol, or phospholipids [10]. Post-heparin lipolytic activity is significantly greater in the young (mean age of 26) than in the elderly (mean age 86) [11].

The decline in lipid clearance might enhance arterial plaque formation, and raises the possibility that the age-associated decrease in endogenous heparin and other proteoglycans contributes to atherogenesis. Hyman Engelberg (1913−2005), a Los Angeles cardiologist, was an early proponent of the concept that atherosclerosis is a heparin deficiency disease. Between 1952 and 2004, Engelberg published nearly 100 papers relating to heparin, lipids and atherosclerosis. He recognized that circulating LPL activity was due to release of lipase bound to the vascular endothelial surface, and noted that more enzyme is released by healthy individuals than patients with known coronary disease [12]. He reported that LPL requires heparin for its activity; the enzyme is inactive in patients with mutations in regions required for heparin binding [13]. Heparin releases hepatic triglyceride lipase as well, contributing to the clearance of chylomicrons and VLDL remnant particles by the liver [14]. Heparin also blocks LPL from cooperating with the alpha-2 macroglobulin receptor in the uptake of LDL particles by atheromatous lesions [15].

Defense of the endothelium

Engelberg was also interested in the role of heparin in safeguarding the endothelium [12]. He reported that heparin reduces the levels of injurious free oxygen radicals by decreasing their release from activated neutrophils, scavenges radicals released by myeloperoxidase, and enhances the activity of the antioxidant, superoxide dismutase. Heparin decreases inflammation by binding cationic proteins, histamine, and activated complement, and reduces free radical release from re-perfused ischemic kidneys [16]. Another source of vascular injury is herpes simplex virus-1 (HSV-1). This virus, as well as cytomegalovirus (CMV), are prevented from being adsorbed to the host cell surface by heparin, which forms an active anti-viral complex with interferon gamma (IFN-γ). In addition, heparin suppresses the production and release of the vasoconstrictor, endothelin, by binding to thrombin, angiotensin II, and arginine vasopressin, shifting the balance in vascular tone from endothelin to the vasodilator, nitric oxide [17]. Lastly, heparin contributes to vessel healing by enhancing the mitogenic activity of vascular endothelial growth factor.

Effects on cell proliferation

In a landmark study, Clowes and Karnowsky reported that heparin suppressed intimal smooth muscle cell proliferation in a rat model of arterial endothelial injury [18]. They speculated that the acidic heparin might bind to and inhibit a basic platelet mitogenic protein. In subsequent studies, Guyton et al [19] reported that the administration of non-anticoagulant heparin, as well as wild-type heparin, inhibited myointimal growth in a rat carotid artery injury model. Further analyses using cultured smooth muscle cells revealed that exposure to heparin, 100 μg/mL for 48 hours, prevented cells from entering S phase and replicating [20]. In other work, heparin inhibited the outgrowth of vascular smooth muscle cells from human coronary artery explants, maintaining the cells in a quiescent state [21]. Finally, vascular smooth muscle cells from veins with restenotic lesions following coronary artery bypass procedures were much less sensitive to growth inhibition by heparin than cells from unaffected veins [22]. This heparin resistance might be due to mitogenic factors in the coronary circulation of patients susceptible to restenosis.

Heparin completely inhibits expression of the c-Myb proto-oncogene, a member of the MYB (myeloblastosis) family of transcription factors that have an essential role in hematopoiesis. It is a downstream target of platelet derived growth factor (PDGF), mediating smooth muscle cell survival [23]. Heparin binds to high affinity receptors on smooth muscle cells and is internalized; [24] in order to block c-Myb expression and inhibit DNA synthesis, heparin requires the presence of transforming growth factor-β (TGF-β) [25]. In fact, much of the anti-proliferative effect of heparin is likely attributable to its potentiation of TGF-β [26].

Heparin has profound effects on a number of mitogens. It blocks PDGF from binding to its cell surface receptor [27], and inhibits the incorporation thrombospondin into the extracellular matrix, preventing it from enhancing the mitogenic action of epidermal growth factor (EGF) on smooth muscle cells. Both EGF and basic fibroblast growth factor (bFGF) require heparin in order to bind to cell surface receptors in the absence of heparan sulfate proteoglycans [28,29]. Vessel pericyte and smooth muscle cell proliferation are dose-dependently inhibited by heparin [30]. From a physiological viewpoint, it might be considered that endothelial cell and smooth muscle cell heparan sulfates constitute a natural blood vessel wall anti-proliferative pathway [31]. Cell proliferation is enhanced by the degradation of heparan by heparanase, an activity that heparin blocks by inhibiting heparanase.

The growth of neoplasms is also influenced by heparin. Moses Judah Folkman (1933–2008) was surgeon-in-chief at Boston Children's Hospital and a prolific investigator. He observed that tumor growth was dependent on the production of feeder blood vessels, and hypothesized that tumors released factors that stimulated vessel growth (angiogenic factors) [32]. He suspected that one of these factors might be heparin because mast cells were conspicuously present at tumor sites and angiogenic activity was inhibited by protamine [33,34]. Although heparin did indeed stimulate angiogenesis, when combined with corticosteroids it was a potent inhibitor of capillary proliferation and caused tumors to regress [35]. In fact, Folkman discovered that a cyclodextrin had >100 times the inhibitory activity of heparin in a chick embryo bioassay [36]. During his long and stellar career, Folkman published more than 400 scientific papers, received many awards including the Wolf Prize in 1992 for his work on angiogenesis, and was selected for membership in the National Academy of Sciences and Institute of Medicine.

Studies by others using an ex-vivo model confirmed that unfractionated heparin stimulates angiogenesis and low molecular weight heparins are inhibitory [37]. The inhibition of endothelial cell proliferation required a chain length of >8 saccharide units, with maximal inhibition by molecules with a molecular weight of 6 kDa [38]. Folkman's concept of treating cancer by inhibiting angiogenesis found support in the development and licensing in 2004 of bevacizumab, an inhibitor of vascular endothelial growth factor [39].

Several years later, Borsig et al. [40] found that sulfated hexasaccharides abrogated tumor interaction with the vascular endothelium by inhibiting P-selectin. More recently, a sulfated, non-saccharide mimetic of heparin hexasaccharide was reported to induced apoptosis of colon cancer stem cells and the self-renewal of tumor xenografts [41]. This work represents an important advance in the development of cancer chemotherapeutic agents.

Inhibition of heparan sulfate cleavage by heparanase

Heparanase is an endoglycosidase that is produced by the inflammatory cells associated with atheromatous plaques in coronary and carotid vessels; it degrades the vessel wall proteoglycan, heparan sulfate [42–44]. Loss of heparan sulfate is associated with occlusive vascular disease, neuroinflammation, sepsis-associated lung injury, and inflammatory bowel disease [45,46]. Also, the degradation of heparan in the extracellular matrix of the kidney contributes to the development of glomerular disease [47]. Heparanase expression is enhanced in tumors and shows an association with tumor size, tumor angiogenesis, and the risk of metastasis [48]. The enzyme has domains that bind both heparin and heparan sulfate; binding of heparin to these domains inhibits the ability of heparanase to cleave heparan sulfate [49]. Inhibition of heparanase might account for some of the anti-proliferative activities as well as anti-inflammatory activities of heparin.

Retardation of inflammation

In 1975, it was reported that the addition of heparin to a suspension of T-lymphocytes decreased rosette formation in a dose-dependent manner [50]. Subsequently, it was found that low concentrations in animal models (5 µg *in vivo*, 0.1 µg/mL *in vitro*) inhibited the expression of T-lymphocyte heparanase and the ability of lymphocytes to migrate and produce a delayed-type hypersensitivity reaction [51]. In these models, heparin suppressed experimental autoimmune diseases and prolonged allograft survival [52].

Heparin in subclinical doses inhibits the adhesion of L- and P-selectins to the surface of leukocytes [53] and in a mouse model prevents the influx of neutrophils at sites of inflammation [54]. It also prevents monocyte adhesion to endovascular metal stents placed in rabbit iliac arteries, and decreases intimal cell proliferation and thickening [55]. Phospholipase A_2 and platelet activating factor (PAF) are mediators of inflammation. Heparin inhibits phospholipase A_2 activity and blocks the airway hyperresponsiveness and pulmonary cell infiltration induced by PAF in the rabbit [56,57].

Heparin has multiple interactions with the complement system, binding to 13 different components [58]. These include members of the classical pathway (C1, C1q, C1 inhibitor, C2, C4, C4b and C4bp), the lectin pathway (mannan-binding lectin-associated serine proteases 1 and 2), and the alternate pathway. Heparin blocks the C3b binding site for factor B and increases the affinity of factor H for C3b, but at low concentrations activates complement by potentiating the alternate pathway amplification cycle in the fluid phase [59]. Vitronectin has a heparin-binding site, enabling it to bind to heparan on the cell surface and be in position to inhibit complement-mediated cell lysis [60]. *In vivo*, heparin infusions in a guinea pig model are associated with inhibition of total hemolytic complement and C3 [61], consistent with its overall anti-inflammatory activity.

Effects on bone

Early investigators were impressed by the effects of heparin on hemostasis and lipids, and speculated that long-term use of the anticoagulant might retard the development of atherothrombotic disease. They soon discovered that daily injections of the drug for many months led to symptomatic fractures that could involve vertebrae as well as other bones [62]. A study in rats revealed that heparin decreases bone formation and increases the rate of bone resorption, the latter by enhancing IL-11 formation of STAT3-DNA complexes and induction of receptor activator of nuclear factor-κB ligand and glycoprotein 130 [63,64]. Osteoblasts decline by 37% and osteoclasts increase by 43%, and the heparin remains sequestered within bone long after dosing is discontinued [65].

Heparin has been given for up to 34 weeks for the prevention and treatment of deep vein thrombosis during pregnancy. Studies show that bone mineral density will decline in up to a third of these women, and 2%−3% will experience fractures [65]. Decreases in bone mineral density are still detectable three years post-partum [66]. Therefore, the long-term administration of unfractionated heparin is generally to be avoided.

Decrease of blood viscosity

Blood viscosity at low shear rates is significantly decreased 4−6 hours after the subcutaneous injection of 5000 U of heparin [67], and 5000 U intravenously decreased plasma but to a lesser extent, whole blood viscosity in patients with an elevated viscosity at baseline [68]. Heparin failed to reduce red cell aggregation as detected by spontaneous echocardiographic contrast formation in an *in vitro* model of nonlaminar blood flow [69]. While heparin does not decrease red blood cell or platelet aggregation, it does bind to a number of plasma proteins (among them, vitronectin, fibronectin, and histidine-rich glycoprotein) and the formation of complexes between heparin and these proteins might account for the reduction in plasma viscosity.

Suppression of aldosterone

Heparin induces potassium retention and enhances the urinary excretion of sodium and chloride by suppressing the secretion of aldosterone [70,71]. Daily subcutaneous injections during a 4-year period were reported to result in hypoaldosteronism, characterized by thinning of the zona glomerulosa of the adrenal cortex and a reduction in the number and affinity of the angiotensin-II receptors [72,73]. Aldosterone suppression occurs as soon as five days after beginning heparin, even with doses as low as 5000 U twice daily. Potassium levels generally increase by an average of 0.4 mEq/L, but frank hyperkalemia occurs in only 7% of patients. Nevertheless, individuals receiving heparin for longer than 3 days should have their potassium levels checked, and in those with other risk factors for hyperkalemia, monitoring should be every 4 days [73]. In addition, clinicians should consider that a rise in potassium levels in an anticoagulated patient might be due to heparin-induced hypoaldosteronism rather than renal failure or other serious disorders.

Non-thrombotic disorders potentially amenable to treatment with heparins

Disorders with preliminary evidence that heparins might be beneficial are described below and listed in Table 2.1.

Acute lung injury and acute respiratory distress syndrome (ARDS) are associated with inflammation, endothelial cell injury, enhanced coagulation, and alveolar fibrin deposition [74]. Studies in rats find that low molecular

TABLE 2.1 Potentially beneficial effects of heparins in non-thrombotic disorders.

Disorder	Comment
Acute lung injury and acute respiratory distress syndrome	Promising animal data and anecdotal reports that nebulized heparin improves survival in people with smoke inhalation
Allergic rhinitis	Small studies showed transient benefit in some patients
Anemia of chronic inflammation	Heparins inhibit expression of hepcidin, potentially increasing iron bioavailability for hemoglobin synthesis
Asthma	Limited experience shows reduction in inflammatory markers
Chronic obstructive pulmonary disease	Significant benefit in small studies of patients with moderate to severe chronic obstructive pulmonary disease; role in cystic fibrosis requires further study.
Cancer	Decreased thromboembolism but no effect on survival; several new agents under active investigation
Hyperlipemia	Safe & effective for hypertriglyceridemic pancreatitis
Inflammatory bowel diseases	Oral formulations of low molecular weight heparins might induce remission of bowel inflammation
Preeclampsia	Stimulate angiogenesis and decrease inflammatory markers
Sepsis	Beneficial effects might be due to thrombus prevention
Tissue repair and wound healing	Experimental work suggests incorporating heparin into growth factor delivery systems accelerates healing.

weight heparin attenuates systemic inflammation and improves survival in endotoxin-induced acute lung injury [75], but a retrospective propensity-matched cohort study of patients with acute lung injury found that treatment with intravenous unfractionated heparin was not associated with reduced mortality [76]. On the other hand, two systemic reviews of published clinical reports note that the use of nebulized heparin improves the survival of patients with smoke inhalation-induced acute lung injury [77,78]. Prospective studies are needed to confirm the benefits of inhaled heparin prior to the general acceptance of this treatment for people with smoke inhalation.

Allergic rhinitis is mediated by eosinophils and mast cells, which interact to form an allergic effector unit [79]. In patients with allergic rhinitis, pre-treatment with intranasal heparin prior to nasal challenge reduced eosinophil influx and the amount of eosinophil cationic protein as well as significantly attenuated mast cell release of histamine and tryptase [80,81]. Although heparin transiently improved rhinitis symptoms, the specific mechanisms responsible for its beneficial effects are unclear and studies involving larger numbers of patients have not been reported.

Anemia of chronic inflammation is characterized by IL-6-stimulated production of hepcidin, a hepatic peptide that restricts iron availability for hemoglobin synthesis. Heparins with molecular weights >4 kDa and 2O- and 6O-sulfation reduce hepatic hepcidin expression by binding to bone morphogenic protein-6 (BMP-6) and interfering with the activation of the BMP/SMAD pathway [82]. Inhibiting hepcidin expression might increase iron bioavailability and mitigate the anemia of chronic inflammation; clinical studies to test this possibility are in progress.

Asthma attacks induced by house-dust mite extract were partially abrogated by heparin given by nebulizer prior to exposure to the allergens, and immediate and late increases in forced expiratory volumes (FEV_1) were demonstrated [83,84]. These salutary responses are probably mediated by heparin's ability to inhibit histamine release by airway mast cells and constrain eosinophilic migration into the bronchi. In addition, heparin decreases the proliferation of cultured human airway smooth muscle cells; this anti-proliferative action is shared with low molecular weight heparin and is independent of its anticoagulant activity [85]. A recent systemic review described seven randomized, controlled, crossover trials involving sample sizes of up to 25 patients, with consistent benefit [86]. Lastly, a study presented in abstract form described 24 patients with asthma given low molecular weight heparin nebulization, 5000−10,000 U twice daily for 14 days [87]. Significant increases in FEV_1 and decreases in eosinophils and lymphocytes in bronchial lavage fluid were reported. Bleeding or other adverse effects appear infrequent, but heparin analogs without anticoagulant activity would be preferred.

Chronic Obstructive Pulmonary Disease (COPD) is associated with chronic bronchial airway inflammation, fibrosis, and mucous hypersecretion resulting in small airway obstruction [88]. During exacerbations of COPD, the secretion of interleukin-8 (IL-8) by endothelial cells drives recruitment of neutrophils, formation of neutrophil and eosinophilic extracellular traps, and release of neutrophil elastase, all of which contribute to airway inflammation. Heparin inhibits IL-8, neutrophil activation, and elastase secretion, and therefore might limit COPD exacerbations. A randomized, placebo-controlled trial of nebulized heparin, 150,000 IU twice daily for 21 days in 9 patients with COPD, showed significantly increased FEV_1 by day 7 [89]. In addition, there were improvements in exercise capacity and dyspnea, and no evidence of systemic anticoagulation. Small, randomized, open-label studies of low molecular weight heparins have reported modest improvements, but do not appear as beneficial as unfractionated heparin.

Cystic Fibrosis: Neutrophil elastase is also a mediator of inflammation in patients with cystic fibrosis, and is inhibited by a 2-O, 3-O-desulfated heparin (ODSH) [90]. Although a clinical trial of inhaled unfractionated heparin, 50,000 IU twice daily, failed to improve sputum clearance or inflammatory markers in 14 adults with cystic fibrosis [91], intra-tracheal ODSH showed anti-protease activity in a murine model of neutrophil elastase-induced airway inflammation [90]. However, ODSH was capable of inhibiting neutrophil elastase in cystic fibrosis sputum only if desoxyribonuclease 1 (dornase) was present [92]. Further studies are needed to determine whether non-anticoagulant heparins will have a role in the treatment of cystic fibrosis.

Cancer: The anti-proliferative activity of heparin has led to clinical trials testing its ability to increase survival in patients with various cancers. Improved outcomes were noted in individuals with early stage small cell lung cancer; those randomized to subcutaneous heparin for 5 weeks had improved response rates (37% vs 23%, P = 0.004) and better median survival (317 days vs 261 days, P = 0.01) [93]. These results encouraged additional trials of heparin in ambulatory cancer patients, but the outcomes have been less satisfactory, and a recently updated Cochrane Systematic Review of 18 randomized trials including more than 9000 patients found that heparin had no effect on mortality at either 12 or 24 months [94]. Although there was a reduced risk of symptomatic venous thromboembolism, there was more bleeding. It is possible that the effects of heparin on cell proliferation and angiogenesis described previously might have more of an impact on some tumor types than others. For example, heparin has been shown to inhibit tumor growth, angiogenesis, and the uptake of extracellular vesicles by malignant gliomas [95]. Unfortunately, the small survival benefit recorded was counterbalanced by an increased risk of hemorrhage. Non-anticoagulant heparins are undergoing further evaluation as anti-cancer agents, and are discussed in the final section of this Chapter.

Hypertriglyceridemia occasionally provokes acute pancreatitis; heparin combined with insulin releases lipases, decreasing lipid levels and contributing to improved outcomes [96]. A prospective, randomized, controlled trial compared high-volume hemofiltration with the combination of LMWH and insulin in 66 patients with hypertriglyceridemic pancreatitis [97]. Persistent organ failure was more frequent in the hemofiltration group (risk ratio 2.42; 95% confidence interval, 1.15−5.11); other outcomes such as the duration of hospitalization, complications, and mortality, were similar.

Inflammatory bowel disease (IBD), specifically ulcerative colitis, underwent remission when heparin was administered for the treatment of venous thromboembolism [98]. Subsequently, a randomized comparison of intravenous heparin with corticosteroids in 20 patients with IBD observed similar responses in clinical activity, stool frequency, and histopathological grading [99]. Several theories were proposed to explain these beneficial effects, and included heparin's antagonism of the pro-inflammatory actions of thrombin, protection of basic fibroblast growth factor, and immunomodulatory activity [100,101]. Additional clinical trials were conducted, but a Cochrane systematic review in 2011 found no evidence to support the use of unfractionated heparin or standard doses of low molecular weight heparins in active ulcerative colitis [102]. However, the higher doses of low molecular weight heparin released by extended colon-release tablets were reported to achieve more frequent clinical remissions and less rectal bleeding, although mucosal healing was not observed [103,104]. Further studies of novel heparin-derived agents should be conducted in this disorder.

Preeclampsia is associated with impaired angiogenesis due to a decrease in pro-angiogenic placental growth factor and an increase in soluble fms-like tyrosine kinase-1 (sFlt1) [105]. Placental growth factor production is stimulated by low dose aspirin, which is currently recommended for preeclampsia prophylaxis. Heparins have been extensively evaluated for preeclampsia prophylaxis [106], but a meta-analysis of eight randomized trials of low molecular weight heparin found no reduction in the risk of pregnancy complications in at-risk women [107]. However, a non-anticoagulant heparin, glycol-split low molecular weight heparin, promoted the secretion of placental growth factor and endothelial cell tube formation from placental villous explants; it also inhibited complement activation and leukocyte adhesion to endothelial cells activated by preeclamptic serum [108]. Clinical evaluation of non-anticoagulant heparins for preeclampsia management is anticipated.

Sepsis During sepsis, the formation of neutrophil extracellular traps is accompanied by the release of cytotoxic histones. These histones can be bound and neutralized by heparin; in a mouse model of sepsis, a non-anticoagulant heparin was shown to bind histones and decrease mortality [109]. The efficacy and safety of heparin in septic patients has been examined by a meta-analysis of nine trials encompassing more than 2600 patients [110]. In six placebo-controlled trials, a small survival benefit was associated with heparin (risk ratio for death, 0.88; 95% CI, 0.77−1.00). However, clinical heterogeneity among the studies was substantial, minor bleeding was increased, and there was no decrease in the duration of mechanical ventilation. A meta-analysis that examined low molecular weight heparin in sepsis treatment showed a reduction in mortality (0.72; 95% CI, 0.57−0.91) but an increase in bleeding events (3.82; 95% CI, 1.81−8.08) [111]. From these analyses, it is not possible to determine whether the improved survivals were due to the anticoagulant or non-anticoagulant activity of the heparins, and their use for the treatment of severe sepsis is still experimental [112].

Tissue repair and wound healing are accompanied by activation and release of growth factors such as Fibroblast Growth Factor-1 (FGF-1), which regulates the proliferation of fibroblasts, blood vessels, and epithelial cells, and FGF-7, responsible for keratinocyte migration and proliferation. Although heparin enhances the stability of FGF-1, it inhibits FGF-7 [113]. On the other hand, introducing heparin at injury sites might stimulate epidermal cell proliferation by releasing epidermal growth factor bound to heparan sulfate [114]. Another option is to deliver growth factors bound to heparin-conjugated nanospheres; employing this approach, La and Yang [115] demonstrated that angiogenesis was accelerated and wound healing enhanced in animal models with full-thickness wounds.

Non-anticoagulant heparins for clinical applications

As noted in Chapter 1, the anticoagulant activity of heparins resides in the pentasaccharide sequence, but this sequence is present on only a third of the saccharide chains. Other biological activities of heparin, such as its anti-inflammatory and anti-proliferative actions, might reside in molecules lacking the pentasaccharide sequence. For example, a disaccharide found in enoxaparin inhibited the release of IL-6 and IL-8 from trypsin-stimulated pulmonary epithelial cells by >70% [116]. Further studies identified a 6-O-sulfated tetrasaccharide fraction that inhibited cytokine release from activated peripheral blood mononuclear cells of allergic asthmatics [117]. Other investigators reported that non-anticoagulant heparin tetrasaccharides decrease binding of neutrophils to cultured cells expressing L- and P-selectins, and reduce neutrophil influx [54]. Fractionated heparins with a chain length as small as 10 saccharides inhibit elastase release, but fail to prevent neutrophil adhesion to stimulated endothelial cells [3]. And heparin rendered non-anticoagulant by O-desulfation still abrogates proliferation of cultured airway smooth muscle cells [85].

Periodate oxidation has been used to prepare non-anticoagulant heparins from unfractionated heparin (NACHs) or LMWH (S-NACHs); hydroxyl groups in the antithrombin binding site are oxidized, removing anticoagulant activity but retaining sulfation [118]. These NACHs are capable of binding to proteins such as hepatocyte growth factor, fibroblast growth factor, and transforming growth factor-β. In addition, S-NACH inhibits P-selectin mediated adhesion of human pancreatic cancer cells to the endothelium of umbilical veins and has anti-metastatic activity in a mouse model of pancreatic cancer [119,120]. Furthermore, it enhanced the anti-tumor activity of gemcitabine in this model.

S-NACH increases the oxygen affinity of hemoglobin S and inhibits the *in vitro* sickling of red blood cells from individuals with sickle cell disease (SCD) [121]. This non-anticoagulant LMWH also decreased circulating levels of sickled red cells in SCD mice, an effect that persisted for up to 6 hours after a dose. In a mouse model of asthma, S-NACH decreased airway eosinophilia, mucus production, and airway hyperresponsiveness, presumably by inhibition of the IL-4/JAK1/STAT6 pathway [122].

N-desulfated heparins inhibit metastasis of human gastric cancers implanted into the stomach of nude mice by blocking expression of basic fibroblast growth factor and tumor angiogenesis, and 2,3-O desulfated heparins hastened acute neurologic recovery in a mouse model of traumatic brain injury by decreasing cerebral leukocyte recruitment, microvascular permeability and edema [123,124]. In addition, N-acetylated, de-O-sulfated heparin derivatives inhibit eosinophil and neutrophil recruitment to the lungs by decreasing P-selectin expression on activated platelets, and abolish pulmonary metastasis by binding to galectin-3 in nude mouse models of melanoma and colon cancer [125,126]. Other research describes a non-anticoagulant heparin prodrug conjugated with a polymeric hydrogel; after local injection, the gel undergoes biodegradation and the sustained release of the heparin inhibits cancer metastasis [127]. Because of their potent anti-inflammatory and anti-proliferative activities, further studies of these non-anticoagulant heparins are anticipated.

Lastly, drugs that inhibit heparanase are being evaluated for their anti-neoplastic activity [128]. PI-88 (Muparfostat) is a fully-sulfated phospho-mannopentaose heparanase mimetic that has been investigated for its ability to prevent recurrences after curative resection of hepatocellular carcinoma [129]. The recurrence-free rate three years after surgery was modestly increased from 50% to 63%; the greatest benefit was observed in patients with elevated levels of heparanase in post-operative plasma samples [130,131]. However, a serious adverse effect of the drug is thrombocytopenia due to the formation of platelet factor-4 antibodies (heparin-induced thrombocytopenia) [132], which might impede further development of this agent. PG545 (Pixatimod) is a heparanase mimetic that binds to the enzyme with high affinity and blocks its catalytic site [133]. Although pre-clinical studies demonstrated reduced proliferation of colon cancer cells and potent lymphoma cell apoptosis [134,135], a phase I clinical trial in 23 patients with advanced solid tumors failed to show objective responses [136]. Another agent under study is SST0001 (Roneparstat), a modified heparin whose ability to inhibit heparanase has been enhanced by glycol-splitting of its nonsulfated uronic acid residues [137]. Roneparstat diminished heparanase-induced shedding of syndecan-1 from myeloma cell lines, suppressed angiogenesis, and reduced the activation of several sarcoma-associated growth factors [138,139]. In preclinical studies, the drug has shown activity against myeloma and pediatric sarcoma xenografts [140]. Necuparanib is also a glycol-split heparin derivative; it has been combined with standard chemotherapeutic agents in a phase I study of 39 patients with metastatic pancreatic carcinoma [141]. A partial response was observed in 38% and 25% achieved stable disease, but skin toxicity (cellulitis) and modest prolongation of the partial thromboplastin time were recorded. Lastly, an ultra-low molecular weight heparin has been synthesized by physico-chemical depolymerization [4]. It produces 70% inhibition of heparanase but has only half the anti-FXa activity of unfractionated heparin, and is being evaluated as an anti-cancer drug. In summary, it appears possible to prepare heparins that are devoid of anticoagulant activity but retain desired anti-inflammatory and anti-proliferative properties. Their safety and efficacy when used alone or in combination with standard agents are still uncertain and await randomized, appropriately-sized clinical trials.

References

[1] Lane DA, Adams L. Non-anticoagulant uses of heparin. N Engl J Med 1993;329:129−30.

[2] Yan Y, Ji Y, Su N, Mei X, Wang Y, Du S, et al. Non-anticoagulant effects of low molecular weight heparins in inflammatory disorders: a review. Carbohydr Polym 2017;160:71−81.

[3] Lever R, Lo WT, Faraidoun M, Amin V, Brown RA, Gallagher J, et al. Size-fractionated heparins have differential effects on human neutrophil function in vitro. Br J Pharmacol 2007;151:837−43.

[4] Achour O, Poupard N, Bridiau N, Bordenave Juchereau S, Sannier F, Piot JM, et al. Anti-heparanase activity of ultra-low-molecular-weight heparin produced by physicochemical depolymerization. Carbohydr Polym 2016;135:316−23.

[5] Yan Y, Guan C, Du S, Zhu W, Ji Y, Su N, et al. Effects of enzymatically depolymerized low molecular weight heparins on CCl_4-induced liver fibrosis. Front Pharmacol 2017;8:514.

[6] Hahn PF. Abolishment of alimentary lipemia following injection of heparin. Science 1943;98:19−20.

[7] Spitzer JJ. Properties of heparin-produced lipemia clearing factor. Am J Physiol 1952;171:492−8.

[8] Borgstrom B, Carlson LA. On the mechanism of the lipolytic action of the lipaemia-clearing factor. Biochim Biophys Acta 1957;24:638−9.

[9] Connor WE, Armstrong ML. Plasma lipoprotein lipase after subcutaneous heparin. Circulation 1961;24:87−93.

[10] Rutstein DD, Castelli WP, Nickerson RJ. Heparin and human lipid metabolism. Lancet 1969;i:1003−8.

[11] Brodows RG, Campbell RG. Effect of age on post-heparin lipase. N Engl J Med 1972;287:969−70.

[12] Engelberg H. Actions of heparin in the atherosclerotic process. Pharmacol Rev 1996;48:327–52.

[13] Beg OU, Meng MS, Skarlatos SI, Previato L, Brunzell JD, Brewer Jr HB, et al. Lipoprotein lipase Bethesda: a single amino acid substitution (Ala-176---Thr) leads to abnormal heparin binding and loss of enzymic activity. Proc Natl Acad Sci USA 1990;87:3474–8.

[14] Busch SJ, Martin GA, Barnhart RL, Jackson RL. Heparin induces the expression of hepatic triglyceride lipase in a human hepatoma (HepG2) cell line. J Biol Chem 1989;264:9527–32.

[15] Larnkjaer A, Nykjaer A, Olivecrona G, Thogersen H, Ostergaard PB. Structure of heparin fragments with high affinity for lipoprotein lipase and inhibitor of lipoprotein lipase binding to alpha 2-macroglobulin-receptor/low −density −lipoprotein-receptor-related protein by heparin fragments. Biochem J 1995;307:205–14.

[16] Nilsson UA, Haraldsson G, Bratell S, Sørensen V, Akerlund S, Pettersson S, et al. ESR-measurement of oxygen radicals in vivo after renal ischaemia in the rabbit. Effects of pre-treatment with superoxide dismutase and heparin. Acta Physiol Scand 1993;147:263–70.

[17] Yokokawa K, Tahara H, Kohno M, Mandal AK, Yanagisawa M, Takeda T. Heparin regulates endothelin production through endothelium-derived nitric oxide in human endothelial cells. J Clin Invest 1993;92:2080–5.

[18] Clowes AW, Karnowsky MJ. Suppression by heparin of smooth muscle cell proliferation in injured arteries. Nature 1977;265:625–6.

[19] Guyton JR, Rosenberg RD, Clowes AW, Karnovsky MJ. Inhibition of rat arterial smooth muscle cell proliferation by heparin. Circ Res 1980;46:625–34.

[20] Reilly CF, Kindy MS, Brown KE, Rosenberg RD, Sonenshein GE. Heparin prevents vascular smooth muscle cell progression through the G_1 phase of the cell cycle. J Biol Chem 1989;264:6990–5.

[21] Caplice NM, West MJ, Campbell GR, Campbell J. Inhibition of human vascular smooth muscle cell growth by heparin. Lancet 1994;344:97–8.

[22] Chan P, Patel M, Betteridge L, Munro E, Schachter M, Wolfe J, et al. Abnormal growth regulation of vascular smooth muscle cells by heparin in patients with restenosis. Lancet 1993;341:341–2.

[23] Chen Y, Xu H, Liu J, Zhang C, Leutz A, Mo X. The c-Myb functions as a downstream target of PDGF-mediated survival signal in vascular smooth muscle cells. Biochem Biophys Res Commun 2007;360:433–6.

[24] Castellot Jr JJ, Wong K, Herman B, Hoover RL, Albertini DF, Wright TC, et al. Binding and internalization of heparin by vascular smooth muscle cells. J Cell Physiol 1985;124:13–20.

[25] McCaffrey TA, Falcone DJ, Brayton CF, Agarwal LA, Welt FG, Weksler BB. Transforming growth factor-beta activity is potentiated by heparin via dissociation of the transforming growth factor-beta/alpha2-macroblobulin inactive complex. J Cell Biol 1989;109:441–8.

[26] Grainger DJ, Witchell CM, Watson JV, Metcalfe JC, Weissberg PL. Heparin decreases the rate of proliferation of rat vascular smooth muscle cells by releasing transforming growth factor beta-like activity from serum. Cardiovasc Res 1993;27:2238–47.

[27] Fager G, Camejo G, Bondjers G. Heparin-like glycosaminoglycans influence growth and phenotype of human arterial smooth muscle cells in vitro. I. Evidence for reversible binding and inactivation of the platelet-derived growth factor by heparin. In Vitro Cell Dev Biol 1992;28A(3 Pt 1):168–75.

[28] Ornitz DM, Yayon A, Flanagan JG, Svahn CM, Levi E, Leder P. Heparin is required for cell-free binding of basic fibroblast growth factor to a soluble receptor and for mitogenesis in whole cells. Mol Cell Biol 1992;12:240−7.

[29] Higashiyama S, Abraham JA, Miller J, Fiddes JC, Klagsbrun M. A heparin-binding growth factor secreted by macrophage-like cells that is related to EGF. Science 1991;251:936−9.

[30] Orlidge A, D'Amore PA. Cell specific effects of glycosaminoglycans on the attachment and proliferation of vascular wall components. Microvasc Res 1986;31:41−53.

[31] Rosenberg RD. Vascular smooth muscle cell proliferation: basic investigations and new therapeutic approaches. Thromb Haemost 1993;70:10−16.

[32] Folkman J. Tumor angiogenesis: therapeutic implications. N Engl J Med 1971;285:1182−6.

[33] Taylor S, Folkman J. Protamine is an inhibitor of angiogenesis. Nature 1982;297:307−12.

[34] Folkman J, Taylor S, Spillberg C. The role of heparin in angiogenesis. Ciba Found Symp 1983;100:132−49.

[35] Folkman J, Langer R, Linhardt RJ, Haudenschild C, Taylor S. Angiogenesis inhibition and tumor regression caused by heparin or a heparin fragment in the presence of corti-sone. Science 1983;221:719−25.

[36] Folkman J, Weisz PB, Joullie MM, Li WW, Ewing WR. Control of antiogenesis with synthetic heparin substitutes. Science 1989;243:1490−3.

[37] Norrby K. Heparin and angiogenesis: a low-molecular-weight fraction inhibits and a high-molecular-weight fraction stimulates angiogenesis systemically. Haemostasis 1993;23 (suppl 1):141−9.

[38] Khorana AA, Sahni A, Altland OD, Francis CW. Heparin inhibition of endothelial cell proliferation and organization is dependent on molecular weight. Arterioscler Thromb Vasc Biol 2003;23:2110−15.

[39] Ribatti D. Napoleone Ferrara and the saga of vascular endothelial growth factor. Endothelium 2008;15:1−8.

[40] Borsig L, Vlodavsky I, Ishai-Michaeli R, Torri G, Vismara E. Sulfated hexasaccharides attenuate metastasis by inhibition of P-selectin and heparanase. Neoplasia 2011;13:445−52.

[41] Boothello RS, Patel NJ, Sharon C, Abdelfadiel EI, Morla S, Brophy DF, et al. A unique non-saccharide mimetic of heparin hexasaccharide inhibits colon cancer stem cells via p38 MAP kinase activation. Mol Cancer Ther 2018; pii: molcanther.0104.2018.

[42] Stevens RL, Colombo M, Gonzales JJ, Hollander W, Schmid K. The glycosaminoglycans of the human artery and their changes in atherosclerosis. J Clin Invest 1976;58:470−81.

[43] Blich M, Golan A, Arvatz G, Sebbag A, Shafat I, Sabo E, et al. Macrophage activation by heparanase is mediated by TLR-2 and TLR-4 and associates with plaque progression. Arterioscler Thromb Vasc Biol 2013;33:e56−65.

[44] Osterholm C, Folkersen L, Lengquist M, Ponten F, Renne T, Li J, et al. Increased expression of heparanase in symptomatic carotid atherosclerosis. Atherosclerosis 2013;226:67−73.

[45] Vlodavsky I, Blich M, Li JP, Sanderson RD, Ilan N. Involvement of heparanase in athero-sclerosis and other vessel wall pathologies. Matrix Biol 2013;32:241−51.

[46] Goldberg R, Meirovitz A, Hirshoren N, Bulvik R, Binder A, Rubinstein AM, et al. Versatile role of heparanase in inflammation. Matrix Biol 2013;32:234−40.

[47] Rabelink TJ, van den Berg BM, Garsen M, Wang G, Elkin M, van der Vlag J. Heparanase: roles in cell survival, extracellular matrix remodeling and the development of kidney disease. Nat Rev Nephrol 2017;13:201−12.

[48] Vlodavsky I, Singh P, Boyango I, Gutter-Kapon L, Elkin M, Sanderson RD, et al. Heparanase: from basic research to therapeutic applications in cancer and inflammation. Drug Resist Update 2016;29:54−75.

[49] Levy-Adam F, Abboud-Jarrous G, Guerrini M, Beccati D, Vlodavasky I, Ilan N. Identification and characterization of heparin/heparan sulfate binding domains of the endoglycosidase heparanase. J Biol Chem 2005;280:20456−66.

[50] Hadfield TL, Marcus S, Smart CR. Heparin and T cells. N Engl J Med 1975;293:1101−2.

[51] Lider O, Mekori YA, Miller T, Bar-Tana R, Vlodavsky I, Baharav E, et al. Inhibition of T lymphocyte heparanase by heparin prevents T cell migration and T cell-mediated immunity. Eur J Immunol 1990;20:493−9.

[52] Lider O, Baharav E, Mekori YA, Miller T, Naparstek Y, Vlodavsky I, et al. Suppression of experimental autoimmune diseases and prolongation of allograft survival by treatment of animals with low doses of heparins. J Clin Invest 1989;83:752−6.

[53] Koenig A, Norgard-Sumnicht K, Linhardt R, Varki A. Differential interactions of heparin and heparin sulfate glycosaminoglycans with the selectins. Implications for the use of unfractionated and low molecular weight heparins as therapeutic agents. J Clin Invest 1998;101:877−89.

[54] Nelson RM, Cecconi O, Roberts WG, Aruffo A, Linhardt RJ, Bevilacqua MP. Heparin oligosaccharides bind L- and P-selectin and inhibit acute inflammation. Blood 1993;82:3253−8.

[55] Rogers C, Welt FGP, Karnovsky MJ, Edelman ER. Monocyte recruitment and neointimal hyperplasia in rabbits. Coupled inhibitory effects of heparin. Arterioscler Thromb Vasc Biol 1996;16:1312−18.

[56] Diccianni MB, Mistry MJ, Hug K, Harmony JA. Inhibition of phospholipase A2 by heparin. Biochim Biophys Acta 1990;1046:242−8.

[57] Sasaki M, Herd CM, Page CP. Effect of heparin and a low-molecular weight heparinoid on PAF-induced airway responses in neonatally immunized rabbits. Br J Pharmacol 1993;.

[58] Sahu A, Panoburn MK. Identification of multiple sites of interaction between heparin and the complement system. Mol Immunol 1993;30:679−84.

[59] Keil LB, Jimenez E, Guma M, Reyes MD, Liguori C, DeBari VA. Biphasic response of complement to heparin: fluid-phase generation of neoantigens in human serum and in a reconstituted alternative pathway amplification cycle. Am J Hematol 1995;50:254−62.

[60] Zaferani A, Talsma D, Richter MKS, Daha MR, Navis GJ, Seelen MA, et al. Heparin/heparan sulphate interactions with complement—a possible target for reduction of renal function loss? Nephrol Dial Transplant 2014;29:515−22.

[61] Weiler JM, Edens RE, Linhardt RJ, Kapelanski DP. Heparin and modified heparin inhibit complement activation in vivo. J Immunol 1992;148:3210−15.

[62] Griffith GC, Nichols Jr G, Asher JD, Flanagan B. Heparin osteoporosis. JAMA 1965;193:91−4.

[63] Muir JM, Andrew M, Hirsh J, Weitz JI, Young E, Deschamps P, et al. Histomorphometric analysis of the effects of standard heparin on trabecular bone in vivo. Blood 1996;88:1314−20.

[64] Walton KJ, Duncan JM, Deschamps P, Shaughnessy SG. Heparin acts synergistically with interleukin-11 to induce STAT3 activation and in vitro osteoclast formation. Blood 2002;100:2530−6.

[65]   Rajgopal R, Bear M, Butcher MK, Shaughnessy SG. The effects of heparin and low molecular weight heparins on bone. Thromb Res 2008;122:293−8.

[66]   Pettila V, Leinonen P, Markkola A, Hiilesmaa V, Kaaja R. Postpartum bone mineral density in women treated for thromboprophylaxis with unfractionated heparin or LMWheparin. Thromb Haemost 2002;87:182−96.

[67]   Erdi A, Thomas DP, Kakkar VV, Lane DA. Effect of low-dose subcutaneous heparin on whole-blood viscosity. Lancet 1976;ii:342−4.

[68]   Ruggiero HA, Castellanos H, Caprissi LF, Caprissi ES. Heparin effect on blood viscosity. Clin Cardiol 1982;5:215−8.

[69]   Fatkin D, Loupas T, Low J, Feneley M. Inhibition of red cell aggregation prevents spontaneous echocardiographic contrast formation in human blood. Circulation 1997;96:889−96.

[70]   Schlatmann RJAFM, Jansen AP, Prenen H, van der Korst JK, Majoor CLH. The natriuretic and aldosterone-suppressive action of heparin and some related polysulfated polysaccharides. J Clin Endocrinol Metab 1964;24:35−47.

[71]   Majoor CLH. Aldosterone suppression by heparin. N Engl J Med 1968;279:1172−3.

[72]   Wilson ID, Goetz FC. Selective hypoaldosteronism after prolonged heparin administration: a case report with postmortem findings. Am J Med 1964;36:635−40.

[73]   Oster JR, Singer I, Fishman LM. Heparin-induced aldosterone suppression and hyperkalemia. Am J Med 1995;98:575−86.

[74]   MacLaren R, Stringer KA. Emerging role of anticoagulants and fibrinolytics in the treatment of acute respiratory distress syndrome. Pharmacotherapy 2007;27:860−73.

[75]   Luan ZG, Naranpurev M, Ma XC. Treatment of low molecular weight heparin inhibits systemic inflammation and prevents endotoxin-induced acute lung injury in rats. Inflammation 2014;37:924−32.

[76]   Hofstra JJ, Vlaar AP, Prins DJ, Koh G, Levi M, Schultz MJ, et al. Early intravenous unfractionated heparin and outcome in acute lung injury and acute respiratory distress syndrome: a retrospective propensity matched cohort study. BMC Pulm Med 2012;12:43.

[77]   Tuinman PR, Dixon B, Levi M, Juffermans NP, Schultz MJ. Nebulized anticoagulants for acute lung injury-a systematic review of preclinical and clinical investigations. Crit Care 2012;16:R70.

[78]   Miller AC, Elamin EM, Suffredini AF. Inhaled anticoagulation regimens for the treatment of smoke inhalation-associated acute lung injury: a systematic review. Crit Care Med 2014;42:413−19.

[79]   Minai-Fleminger Y, Elishmereni M, Vita F, Soranzo MR, Mankuta D, Zabucchi G, Levi-Schaffer F. Ultrastructural evidence for human mast cell-eosinophil interactions in vitro. Cell Tissue Res 2010;341(3):405−15.

[80]   Vancheri C, Mastruzzo C, Armato F, Tomaselli V, Magri S, Pistorio MP, et al. Intranasal heparin reduces eosinophil recruitment after nasal allergen challenge in patients with allergic rhinitis. J Allergy Clin Immunol 2001;108:703−8.

[81]   Zeng D, Prosperini G, Russo C, Spicuzza L, Cacciola RR, Di Maria GU, et al. Heparin attenuates symptoms and mast cell degranulation induced by AMP nasal provocation. J Allergy Clin Immunol 2004;114:316−20.

[82]   Poli M, Asperti M, Ruzzenenti P, Naggi A, Arosio P. Non-anticoagulant heparins are hepcidin antagonists for the treatment of anemia. Molecules 2017;22 pii: E598.

[83]   Bowler SD, Smith SM, Lavercombe PS. Heparin inhibits the immediate response to antigen in the skin and lungs of allergic subjects. Am Rev Respir Dis 1993;147:160−3.

[84] Diamant Z, Timmers MC, van der Veen H, Page CP, Van der Meer FJ, Sterk PJ. Effect of inhaled heparin on allergen-induced early and late asthmatic responses in patients with atopic asthma. Am J Respir Crit Care Med 1996;153:1790−5.

[85] Kanabar V, Hirst SJ, O'Connor BJ, Page CP. Some structural determinants of the antiproliferative effect of heparin-like molecules on human airway smooth muscle. Br J Pharmacol 2005;146:370−7.

[86] Mousavi S, Moradi M, Khorshidahmad T, Motamedi M. Anti-inflammatory effects of heparin and its derivatives: a systematic review. Adv Pharmacol Sci 2015;2015:507151.

[87] Fal AM, Kraus-Filarska M, Miecielica J, Malolepszy J. Mechanisms of action of nebulized low-molecular-weight heparin in patients with bronchial asthma. J Allergy Clin Immunol 2004;113(2, suppl):S36.

[88] Shute JK, Puxeddu E, Calzetta L. Therapeutic use of heparin and derivatives beyond anticoagulation in patients with bronchial asthma or COPD. Curr Opin Pharmacol 2018;40:39−45.

[89] Shute JK, Calzetta L, Cardaci V, di Toro S, Page CP, Cazzola M. Inhaled nebulised unfractionated heparin improves lung function in moderate to very severe COPD: a pilot study. Pulm Pharmacol Ther 2018;48:88−96.

[90] Griffin KL, Fischer BM, Kummarapurugu AB, Zheng S, Kennedy TP, Rao NV, et al. 2-O, 3-O-desulfated heparin inhibits neutrophil elastase-induced HMGB-1 secretion and airway inflammation. Am J Respir Cell Mol Biol 2014;50:684−9.

[91] Serisier DJ, Shute JK, Hockey PM, Higgins B, Conway J, Carroll MP. Inhaled heparin in cystic fibrosis. Eur Respir J 2006;27:354−8.

[92] Kummarapurugu AB, Afosah DK, Sankaranarayanan NV, Navaz Gangji R, Zheng S, Kennedy T, et al. Molecular principles for heparin oligosaccharide-based inhibition of neutrophil elastase in cystic fibrosis. J Biol Chem 2018;293:12480−90.

[93] Lebeau B, Chastang C, Brechot JM, Capron F, Dautzenberg B, Delaisements C, et al. Subcutaneous heparin treatment increases survival in small cell lung cancer. "Petites Cellules" group. Cancer 1994;74:38−45.

[94] Akl EA, Kahale LA, Hakoum MB, Matar CF, Sperati F, Barba M, et al. Parenteral anticoagulation in ambulatory patients with cancer. Cochrane Database Syst Rev 2017;9:CD006652.

[95] Schnoor R, Maas SLN, Broekman MLD. Heparin in malignant glioma: review of preclinical studies and clinical results. J Neurooncol 2015;124:151−6.

[96] Kuchay MS, Farooqui KJ, Bano T, Khandelwal M, Gill H, Mithal A. Heparin and insulin in the management of hypertriglyceridemia-associated pancreatitis: case series and literature review. Arch Endocrinol Metab 2017;61:198−201.

[97] He WH, Yu M, Zhu Y, Xia L, Liu P, Zeng H, et al. Emergent triglyceride-lowering therapy with early high-volume hemofiltration against low-molecular-weight heparin combined with insulin in hypertriglyceridemic pancreatitis: a prospective randomized controlled trial. J Clin Gastroenterol 2016;50:772−8.

[98] Gaffney PR, Doyle CT, Gaffney A, Hogan J, Hayes DP, Annis P. Paradoxical response to heparin in 10 patients with ulcerative colitis. Am J Gastroenterol 1995;90:220−3.

[99] Ang YS, Mahmud N, While B, Byrne M, Kelly A, Lawler M, et al. Randomized comparison of unfractionated heparin with corticosteroids in severe active inflammatory bowel disease. Aliment Pharmacol Ther 2000;14:1015−22.

[100] White B, Ang YS, Mahmud N, Keeling PWN, Smilth OP. Heparin and inflammatory bowel disease. Lancet 1999;354:1122−3.

[101] Day R, Forbes A. Heparin, cell adhesion, and pathogenesis of inflammatory bowel disease. Lancet 1999;354:62−5.

[102] Chande N, MacDonald JK, Wang JJ, McDonald JW. Unfractionated or low molecular weight heparin for induction of remission in ulcerative colitis: a Cochrane inflammatory bowel disease and functional bowel disorders systematic review of randomized trials. Inflamm Bowel Dis 2011;17:1979−86.

[103] Celasco G, Papa A, Jones R, Moro L, Bozzella R, Surace MM, et al. Clinical trial: oral colon-release parnaparin sodium tablets (CB-01-05 MMX) for active left-sided ulcerative colitis. Aliment Pharmacol Ther 2010;31:375−86.

[104] Baumgart DC. CB-01-05-MMX, a novel oral controlled-release low molecular weight heparin for the potential treatment of ulcerative colitis. Curr Opin Investig Drugs 2010;11:571−6.

[105] Armaly Z, Jadaon JE, Jabbour A, Abassi ZA. Preeclampsia: novel mechanisms and potential therapeutic approaches. Front Physiol 2018;9:973.

[106] Wat JM, Audette MC, Kingdom JC. Molecular actions of heparin and their implications in preventing pre-eclampsia. J Thromb Haemost 2018;16:1510−22.

[107] Rodger MA, Gris JC, de Vries JIP, Martinelli I, Rey É, Schleussner E, et al. Low-molecular-weight heparin and recurrent placenta-mediated pregnancy complications: a meta-analysis of individual patient data from randomised controlled trials. Lancet 2016;388:2629−41.

[108] Wat JM, Hawrylyshyn K, Baczyk D, Greig IR, Kingdom JC. Effects of glycol-split low molecular weight heparin on placental, endothelial, and anti-inflammatory pathways relevant to preeclampsia. Biol Reprod 2018;. Available from: https://doi.org/10.1093/biolre/ioy127.

[109] Wildhagen KC, García de Frutos P, Reutelingsperger CP, Schrijver R, Aresté C, Ortega-Gómez A, et al. Nonanticoagulant heparin prevents histone-mediated cytotoxicity in vitro and improves survival in sepsis. Blood 2014;123:1098−101.

[110] Zarychanski R, Abou-Setta AM, Kanji S, Turgeon AF, Kumar A, Houston DS, et al. The efficacy and safety of heparin in patients with sepsis: a systematic review and metaanalysis. Crit Care Med 2015;43:511−18.

[111] Fan Y, Jiang M, Gong D, Zou C. Efficacy and safety of low-molecular-weight heparin in patients with sepsis: a meta-analysis of randomized controlled trials. Sci Rep 2016;6:25984.

[112] Kalil AC. Should heparin be used to treat patients with severe sepsis? Crit Care Med 2015;43:694−5.

[113] Olczyk P, Mencner Ł, Komosinska-Vassev K. Diverse roles of heparan sulfate and heparin in wound repair. Biomed Res Int 2015;2015:549417.

[114] Lever R, Page CP. Non-anticoagulant effects of heparin: an overview. In: Lever R, Mulloy B, Page CP, editors. Heparin-a century of progress. New York: Springer; 2012. p. 281−305.

[115] La WG, Yang HS. Heparin-conjugated poly (lactic-co-glycolic acid) nanospheres enhance large-wound healing by delivering growth factors in platelet-rich plasma. Artif Organs 2015;39:388−94.

[116] Shastri MD, Stewart N, Horne J, Peterson GM, Gueven N, Sohal SS, et al. In-vitro suppression of IL-6 and IL-8 release from human pulmonary epithelial cells by non-anticoagulant fraction of enoxaparin. PLoS One 2015;10:e0126763.

[117] Shastri MD, Stewart N, Horne J, Zaidi ST, Sohal SS, Peterson GM, et al. Non-anticoagulant fractions of enoxaparin suppress inflammatory cytokine release from

peripheral blood mononuclear cells of allergic asthmatic individuals. PLoS One 2015;10: e0128803.

[118] Ouyang Y, Yu Y, Zhang F, Chen J, Han X, Xia K, et al. Non-anticoagulant low molecular weight heparins for pharmaceutical applications. J Med Chem 2019;62:1067−73.

[119] Sudha T, Phillips P, Kanaan C, Linhardt RJ, Borsig L, Mousa SA. Inhibitory effect of non-anticoagulant heparin (S-NACH) on pancreatic cancer cell adhesion and metastasis in human umbilical cord vessel segment and in mouse model. Clin Exp Metastasis 2012;29:431−9.

[120] Sudha T, Yalcin M, Lin HY, Elmetwally AM, Nazeer T, Arumugam T, et al. Suppression of pancreatic cancer by sulfated non-anticoagulant low molecular weight heparin. Cancer Lett 2014;350:25−33.

[121] Mousa SA, Darwish NHE, Muralidharan-Chari V, Qari M, Qiukan C, Russell JE, et al. Multi-modal mechanisms and anti-sickling of novel sulfated non-anticoagulant low molecular weight heparin in sickle cell disease. ASH Abstract 2018;265.

[122] Ghonim MA, Wang J, Ibba SV, Luu HH, Pyakurel K, Benslimane I, et al. Sulfated non-anticoagulant heparin blocks Th2-induced asthma by modulating the IL-4/signal transducer and activator of transcription 6/Janus kinase 1 pathway. J Transl Med 2018;16:243.

[123] Chen JL, Fan J, Chen MX, Dong Y, Gu JZ. Effect of non-anticoagulant N-desulfated heparin on basic fibroblast growth factor expression, angiogenesis, and metastasis of gastric carcinoma in vitro and in vivo. Gastroenterol Res Pract 2012;2012:752940.

[124] Nagata K, Suto Y, Cognetti J, Browne KD, Kumasaka K, Johnson VE, et al. Early low-anticoagulant desulfated heparin after traumatic brain injury: reduced brain edema and leukocyte mobilization is associated with improved watermaze learning ability weeks after injury. J Trauma Acute Care Surg 2018;84:727−35.

[125] Riffo-Vasquez Y, Somani A, Man F, Amison R, Pitchford S, Page CP. A non-anticoagulant fraction of heparin inhibits leukocyte diapedesis into the lung by an effect on platelets. Am J Respir Cell Mol Biol 2016;55:554−63.

[126] Duckworth CA, Guimond SE, Sindrewicz P, Hughes AJ, French NS, Lian LY, et al. Chemically modified, non-anticoagulant heparin derivatives are potent galectin-3 binding inhibitors and inhibit circulating galectin-3-promoted metastasis. Oncotarget 2015;6:23671−87.

[127] Andrgie AT, Mekuria SL, Addisu KD, Hailemeskel BZ, Hsu WH, Tsai HC, et al. Non-anticoagulant heparin prodrug loaded biodegradable and injectable thermoresponsive hydrogels for enhanced anti-metastasis therapy. Macromol Biosci 2019;19:e1800409.

[128] Vlodavsky I, Singh P, Boyango I, Gutter-Kapon L, Elkin M, Sanderson RD, et al. Heparanase: from basic research to therapeutic applications in cancer and inflammation. Drug Resist Updat 2016;29:54−75.

[129] Liu CJ, Lee PH, Lin DY, Wu CC, Jeng LB, Lin PW, et al. Heparanase inhibitor PI-88 as adjuvant therapy for hepatocellular carcinoma after curative resection: a randomized phase II trial for safety and optimal dosage. J Hepatol 2009;50:958−68.

[130] Liu CJ, Chang J, Lee PH, Lin DY, Wu CC, Jeng LB, et al. Adjuvant heparanase inhibitor PI-88 therapy for hepatocellular carcinoma recurrence. World J Gastroenterol 2014;20:11384−93.

[131] Liao BY, Wang Z, Hu J, Liu WF, Shen ZZ, Zhang X, et al. PI-88 inhibits postoperative recurrence of hepatocellular carcinoma via disrupting the surge of heparanase after liver resection. Tumour Biol 2016;37:2987−98.

[132] Kudchadkar R, Gonzalez R, Lewis KD. PI-88: a novel inhibitor of angiogenesis. Expert Opin Investig Drugs 2008;17:1769−76.

[133] Hammond E, Handley P, Dredge K, Bytheway I. Mechanisms of heparanase inhibition by the heparan sulfate mimetic PG545 and three structural analogues. FEBS Open Bio 2013;3:346−51.

[134] Singh P, Blatt A, Feld S, Zohar Y, Saadi E, Barki-Harrington L, et al. The heparanase inhibitor PG545 attenuates colon cancer initiation and growth, associating with increased p21 expression. Neoplasia 2017;19:175−84.

[135] Weissmann M, Bhattacharya U, Feld S, Hammond E, Ilan N, Vlodavsky I. The heparanase inhibitor PG545 is a potent anti-lymphoma drug: mode of action. Matrix Biol 2018; pii: S0945-053X(18)30174-4.

[136] Dredge K, Brennan TV, Hammond E, Lickliter JD, Lin L, Bampton D, et al. A phase I study of the novel immunomodulatory agent PG545 (pixatimod) in subjects with advanced solid tumours. Br J Cancer 2018;118:1035−41.

[137] Naggi A, Casu B, Perez M, Torri G, Cassinelli G, Penco S, et al. Modulation of the heparanase-inhibiting activity of heparin through selective desulfation, graded N-acetylation, and glycol splitting. J Biol Chem 2005;280:12103−13.

[138] Ritchie JP, Ramani VC, Ren Y, Naggi A, Torri G, Casu B, et al. SST0001, a chemically modified heparin, inhibits myeloma growth and angiogenesis via disruption of the heparanase/syndecan-1 axis. Clin Cancer Res 2011;17:1382−93.

[139] Cassinelli G, Favini E, Dal Bo L, Tortoreto M, De Maglie M, Dagrada G, et al. Antitumor efficacy of the heparan sulfate mimic roneparstat (SST0001) against sarcoma models involves multi-target inhibition of receptor tyrosine kinases. Oncotarget 2016;7:47848−63.

[140] Cassinelli G, Lanzi C, Tortoreto M, Cominetti D, Petrangolini G, Favini E, et al. Antitumor efficacy of the heparanase inhibitor SST0001 alone and in combination with antiangiogenic agents in the treatment of human pediatric sarcoma models. Biochem Pharmacol 2013;85:1424−32.

[141] O'Reilly EM, Roach J, Miller P, Yu KH, Tjan C, Rosano M, et al. Safety, pharmacokinetics, pharmacodynamics, and antitumor activity of necuparanib combined with nab-paclitaxel and gemcitabine in patients with metastatic pancreatic cancer: phase I results. Oncologist 2017;22 1429-e139.

Part II

Anticoagulant Heparins: Clinical Applications

Chapter 3

Prevention of thrombosis

Among the investigators studying heparin in the 1930s, surgeons in particular recognized its potential to treat thrombotic disorders. A tragic and frustrating surgical experience is the successful performance of a difficult operation with minimal blood loss, only to have the patient die shortly postoperatively of a pulmonary embolism. Clarence Crafoord, a Swedish thoracic surgeon, was well-aware of this complication, having performed several pulmonary embolectomies on postoperative patients. He recognized that these thromboses might be averted by systemic anticoagulant therapy, and when heparin became available in 1935 from the Jorpes laboratory, Crafoord and his colleagues began administering it to their postoperative and postpartum patients. No instances of thrombosis were recorded in 657 patients receiving at least 250 mg heparin daily for 5−10 days following operative procedures [1]. Jorpes also records that daily infusions of his heparin restored the visual acuity of a patient with acute central retinal vein thrombosis and resulted in clinical improvement in a patient with posterior inferior cerebellar artery occlusion. At roughly the same time, a Canadian surgeon, GDW Murray, and Best reported that heparin was effective in preventing pulmonary embolism in 315 postoperative patients; in addition, 28 patients with thrombophlebitis and seven with pulmonary embolism were successfully treated [2].

The clinical use of unfractionated heparin (UH) for the prevention of thrombosis was stimulated by an increase in the occurrence of venous thromboembolism (VTE) during World War II. Doctor Keith Simpson, a Lecturer in Forensic Medicine to Guy's Hospital, reported 24 autopsied cases of pulmonary embolism in 1940, as compared to only 4 in 1939 [3]. More than 80% of those dying were either in air-raid shelters or death occurred soon after leaving such facilities. In bomb shelters, people were relatively immobile for prolonged periods. Simpson described one 60 year old woman who sat in a deckchair continuously for 10 hours. In such chairs, the front edge of the seat compresses the leg veins. On arising, she noted leg cramps and swelling; after walking for 8−10 minutes, she collapsed and died. The findings at autopsy were obesity, mild leg varicosities, an anti-mortem clot in the tibial veins, and short clots of small caliber lodged in the first subdivisions of the pulmonary artery. Simpson recommended that obese or elderly

The Heparins. DOI: https://doi.org/10.1016/B978-0-12-818781-4.00003-0

59

persons be provided with bunks so they could be supine rather than sitting: anticoagulant therapy was not available in London at this time.

During World War II, UH could be obtained in Sweden from the laboratory of Doctor Jorpes at the Karolinska Institute. H. Zilliacus reported on the results of five years' treatment of thrombosis at a series of Swedish hospitals between 1940 and 1945 [4]. VTE occurred in 1158 of 256,282 (0.45%), with a mortality rate of 14%. However, during the interval from 1941 to 1944, mortality declined from 8.2% to 5.9%, corresponding to an increase in the use of anticoagulant therapy. Data from a single institution (Mariestad Hospital, Sweden) comparing mortality before and after UH (1929−38 vs 1940−45) showed a 91% decline in mortality and a decrease in fatal embolism from 47 to 3 cases. Zilliacus concluded that specific treatment with UH limited thrombosis to the lower leg, decreased post-thrombotic sequelae, and helped prevent fatal and non-fatal pulmonary embolism.

Surgeons continued to be interested in preventing venous VTE after operations; Vijay Kakkar (1937−2016) made major contributions in this area. He received his medical education in India, continued his training at King's College Hospital in the United Kingdom, and in 1964, became a fellow of the Royal College of Surgeons [5]. His interest in the prophylaxis of venous thrombosis arose after 3-year-old he had operated upon died of multiple pulmonary emboli. In 1972, Kakkar and associates reported that small doses of UH (5000 U) could be given subcutaneously prior to and following surgery without incurring increased intraoperative bleeding or postoperative thrombosis [6]. In fact, pulmonary emboli were recognized in only one of 222 patients receiving prophylaxis. This study led to a larger, prospective, clinical trial that randomized 4121 patients undergoing major surgical procedures [7]. Acute, massive pulmonary emboli occurred significantly more often in patients not receiving prophylactic UH (0.8% vs 0.1%, $P < 0.005$). There was also a significant decrease in the percentage of patients with leg vein thrombi detected by venography (1.5% vs 0.5%) or by ^{125}I-fibrinogen leg scanning (24.6% vs 7.7%). Analysis of postoperative bleeding showed no statistically significant differences in hemoglobin decline or need for blood transfusions, although more heparin-exposed patients developed wound hematomas ($P < 0.01$). The studies of Kakkar demonstrated that low-dose UH was safe and effective in preventing thrombosis in surgical patients, and this therapy was soon extended to medical patients as well. Kakkar established a Thrombosis Research Institute to further investigation into the causes, prevention, and treatment of thrombotic disease. His contributions to improving surgical outcomes were recognized by awards of a Hunterian professorship from the Royal College of Surgeons, a lifetime achievement award from the International Union of Angiology, and, in 2010, the Order of the British Empire.

J.G. Sharnoff was an outspoken proponent for the use of UH to prevent fatal perioperative thrombosis, reporting data from his extensive personal experience [8,9]. In the early 1970s, a controlled clinical trial showed that UH, 5000 IU given subcutaneously 2 hours preoperatively and every 12 hours postoperatively for 7 days, significantly decreased proximal vein thrombosis [10]. Another trial published the following year reported that UH, 5000 IU every 8 hours, decreased the rate of postoperative venous thrombosis from 16.1% to 1.9% without an increase in clinically important bleeding [11]. These early studies stimulated the development of more effective and safer anticoagulants to prevent thrombosis in vulnerable patients.

In recent years, the complications of surgery, including perioperative VTE, have dramatically declined due to improvements in anesthesia, less-invasive procedures, and shorter operating times, as well as the use of antithrombotic agents. This fact has to be taken into consideration when comparisons are made of trials using older versus more recently developed anticoagulants. Nevertheless, low molecular weight heparin (LMWH) has had a profound impact. Edwin W. Salzman (1928−2011), Professor of Surgery at Harvard, a coagulation expert and deputy editor of the New England Journal of Medicine, wrote an editorial in 1986 entitled "Low-Molecular-Weight Heparin: Is Small Beautiful?" [12] He discussed the use of the anticoagulant for preventing thrombosis in patients having total hip replacement, mentioned possible roles in hemodialysis, cardiopulmonary bypass, and the treatment of VTE, and concluded that LMWH was a promising innovation in antithrombotic treatment. A few of these applications of LMWH will be described in the following sections.

Hip arthroplasty and hip fracture

From the clinical observations previously mentioned, early workers recognized that unfractionated heparin was effective in preventing VTE in patients undergoing hip surgery, but there was considerable uncertainty about when to initiate therapy, the optimal dose, and the duration of treatment. Some investigators advocated giving UH pre-operatively; Sharnoff [13] recommended starting the anticoagulant the evening prior to surgery but others gave it 2 hours before operation and continued dosing every 8 hours for 7−10 days [14]. Many gave UH every 12 hours and continued dosing for at least 10−14 days [15]. All showed significant decreases in post-operative thrombosis with no increase in bleeding, but these studies were not randomized or blinded.

A controlled, randomized, double-blind trial that assigned patients to placebo or 5000 U of UH pre-operatively and continued every 8 hours for 7−10 days noted only a small decrease in deep vein thrombosis (DVT) (54% vs 46%, P = non-significant) [16]. However, thrombi were detected in most of the UH group after treatment had been discontinued; during the first 7 post-operative days, DVT was present in 48% of controls and 19% of treated patients. A progressive decline in post-operative clotting times had been described by Sharnoff, who increased the dose of UH to maintain test results within the normal range, a strategy that was effective in preventing late-onset DVTs [17]. A subsequent study by others confirmed that post-operative patients require progressively more UH to keep the activated partial thromboplastin time between 31.5s and 36 s [18]. Adjusting the dose of UH versus giving fixed doses of 3500 U every 8 hours reduced the frequency of DVT from 39% to 13% (P < 0.01) without an increase in bleeding. Lastly, a meta-analysis that pooled data from 27 trial groups encompassing 1745 patients undergoing total hip replacement and receiving various UH regimens calculated that 24% developed DVT [19].

Although it was clear from the trial data that adjusted-dose UH was more effective and safer than fixed-dose UH, most caregivers were unenthusiastic about performing daily clotting assays and hastened to embrace LMWHs because these anticoagulants were usually started postoperatively, were given subcutaneously once or twice daily, and did not require monitoring. In addition, LMWHs were brand-name products requiring randomized, controlled clinical trials to achieve FDA approval, and the pharmaceutical companies that manufactured them were eager to sponsor this research. The earliest trials compared LMWHs with placebo, and later with UH, but infrequently with another LMWH. The introduction of fondaparinux, a synthetic penta-saccharide, led to trials comparing this agent with enoxaparin. LMWH and fondaparinux mainly inhibit factor Xa; a study in patients undergoing hip replacement showed a statistically significant relationship between the anti-factor Xa level and clinical outcomes for thrombosis and wound hematoma [20]. In the following paragraphs, studies with UH, LMWHs, fondaparinux, and direct oral anticoagulants will be summarized. Dalteparin, enoxaparin, and fondaparinux are the heparins currently available in the U.S. for thromboprophylaxis in orthopedic patients; the manufacturer's recommended doses are displayed in Table 3.1.

One of the earliest trials of LMWH randomized 100 patients undergoing hip surgery to **enoxaparin**, 30 mg twice daily, or placebo, and continued for 14 days [21]. DVT was reported in significantly fewer patients assigned to enoxaparin (12% vs 42%, P = 0.0007), with similar rates of bleeding (4% in each group). A dose-finding study reported that smaller percentages of patients receiving 30 mg twice daily or 40 mg once daily had DVTs than those assigned to 10 mg (14% and 11% vs 25%), although the latter had less bleeding (5% vs 13%) [22]. Another study compared enoxaparin, 30 mg every 12 hours with UH, 5000 U every 8 hours [23]. Again, a significantly smaller percentage of patients receiving enoxaparin had DVTs (6% vs 15%, P = 0.03) without an increase in hemorrhages (12% in each group). Furthermore, the mean duration of hospitalization was shorter among patients receiving enoxaparin (9.5 days vs 11.3 days, P = 0.01) [24]. Eventually a double-blind, randomized, controlled trial compared enoxaparin, 30 mg twice daily, to UH, 7500 U twice daily [25]. While the percentage

TABLE 3.1 Manufacturer's recommended doses of low-molecular-weight and synthetic heparins for thromboprophylaxis in orthopedic surgery.

Anticoagulant	Hip replacement	Knee replacement	Hip fracture
Dalteparin	Post-op start: 2500 IU 4−8 h, then 5000 IU daily;	N/A	N/A
	Pre-op start: 2500 IU 2 h before & 2500 IU 4−8 h post-op; then 5000 IU daily;		
	Pre-op start: 5000 IU evening before & 5000 IU 4−8 h post-op; then 5000 IU daily.		
	Continue for up to 14 d[a]		
Enoxaparin	30 mg every 12 h beginning 12−24 h post-op & continuing for up to14 days.	30 mg every 12 h beginning 12−24 h post-op & continuing for up to 14 d[a]	N/A
	40 mg beginning 12 h pre-op, then daily & continuing for up to 14 d[a]		
	Continued prophylaxis: 40 mg daily for 3 wk		
Fondaparinux	2.5 mg daily, beginning 6−8 h post-op & continued for up to 11 d[a]	2.5 mg daily, beginning 6−8 h post-op & continued for up to 11 d[a]	2.5 mg daily, beginning 6−8 h post-op & continued for 32 d

[a]Duration of prophylaxis based on assessment of patient risks for thrombosis or bleeding.
N/A: not a listed indication.

of patients in each group developing DVTs was similar (UH, 23.2% vs enoxaparin, 19.4%), the frequency of bleeding was significantly higher in those receiving UH (9.3% vs 5.1%, P = 0.035). These trials established that enoxaparin, 30 mg twice daily, could replace UH as prophylaxis against venous thrombosis in patients undergoing hip surgery, but there were three issues that needed to be resolved: the timing of the initial dose of LMWH, the duration of treatment, and the cost-effectiveness of enoxaparin compared to UH.

The ideal time to start anticoagulant prophylaxis for hip arthroplasty was unclear; some clinicians began treatment prior to operation and others started therapy 4−24 hours after completion of the procedure. Hull et al. [26] analyzed published randomized trials and concluded that LMWH administered 4−6 hours after surgery was more effective than prophylaxis given 12 hours preoperatively or 12−24 hours post-operatively; giving doses ≤ 2 hours before surgery increased major bleeding. Indeed, most surgeons were concerned that giving an anticoagulant in close proximity to an operation would increase the rate of hemorrhagic complications; they preferred to start prophylaxis 12−24 hours postoperatively, and that is the regimen currently recommended in the enoxaparin package insert.

As previously noted, venous thrombosis and pulmonary emboli, some fatal, occurred as late as 2−3 weeks following hip surgery. To determine the incidence of DVT after leaving the hospital and whether these DVT might be prevented by enoxaparin, investigators recruited patients prior to discharge and randomized them to receive the anticoagulant or placebo. All patients received enoxaparin during hospitalization and were without objective evidence of DVT at discharge. In the study of Bergqvist et al. [27], discharge was at day 11 or 12 and enoxaparin, 40 mg daily, was given for 18−19 days. Thromboembolism was detected in 28% of 233 evaluable patients and was more frequent in patients receiving placebo (39% vs 18%, $P < 0.001$). Planes et al. [28] randomized 173 patients and observed DVTs in 17.3%, of whom 19.3% were receiving placebo and 7.1% enoxaparin ($P = 0.018$). The results of these two trials provide the basis for the current suggestion that enoxaparin, 40 mg daily, be continued for 3 weeks following hospitalization for total hip arthroplasty.

The cost-effectiveness of LMWH compared with UH or warfarin has been examined. Based on a 2.6 to 1 price ratio of LMWH to UH, the use of the former would save the healthcare system about $50,000 per 1000 patients treated, but this would decline to $10,000 per 1000 patients if the price ratio was 10:1 [29]. As compared to warfarin, the total cost per patient of prophylaxis and management of DVT would be $121 more with enoxaparin, but because the LMWH prevented more DVTs, pulmonary emboli, and deaths, the incremental cost-effectiveness of enoxaparin was $29,120 per life-year gained [30]. Another analysis calculated that the cost-effectiveness of enoxaparin would be approximately $12,000 per death averted [31].

A randomized, double-blind trial compared **dalteparin**, 2500 U given 2 hours pre-operatively and 12 hours post-operatively, followed by 5000 U daily, with placebo and reported fewer DVTs in the anticoagulant-treated group (16% vs 35%, $P < 0.02$) [32]. In another trial that compared dalteparin with UH, 5000 U of the LMWH were given 12 hours pre operatively and then daily for 10 days; the UH dose was 5000 U given 2 hours pre-operatively and then every 8 hours [33]. Although the overall rate of thrombosis was similar in the two groups, a significantly higher percentage of patients in the UH group had femoral vein thrombosis (31% vs 10%, P = 0.011) and pulmonary emboli (31% vs 12%, P = 0.016). Furthermore, blood loss and transfusions were greater in those receiving UH. Dalteparin was also compared with warfarin in a trial in which patients received the LMWH, 2500 U within 2 hours pre-operatively and a second dose the night of the operation; subsequently, 5000 U were given daily until patients were assessed for venous thromboembolism (VTE) on day 7 [34]. Significantly fewer patients in the dalteparin group developed DVT (15% vs 26%, P = 0.006), but more had bleeding at the operative site (4% vs 1%, P = 0.03) or required transfusions on days 1−8 postoperatively (48% vs 31%, P < 0.001). Another trial in hip fracture patients compared dalteparin, 2500 U given 2 hours pre-operatively followed by 5000 U daily for 9 days, with UH, 5000 U every 8 hours [35]. VTE was detected in significantly fewer patients receiving UH than dalteparin (14% vs 43%, P = 0.04). Based on the data from these and other studies, dalteparin dosing might begin pre-operatively with 5000 IU the evening prior to surgery followed by 5000 IU 4−8 hours postoperatively, or 2500 IU 2 hours before surgery followed by 2500 IU 4−8 hours after surgery, or started post-operatively with 2500 IU 4−8 hours after surgery. Dosing is continued at 5000 IU daily for at least 7−10 days.

Two trials were performed to examine whether prolonging thromboprophylaxis beyond 1 week was beneficial after hip replacement surgery. In the first study, 265 patients were given 5000 U of dalteparin the evening before surgery and daily for 7 days, then randomized to placebo or dalteparin for the next 4 weeks [36]. In addition, all patients received perioperative infusions of dextran and wore graded elastic stockings for the first week. At day 7, 15.9% of patients had DVTs; at 4 weeks, DVT was detected in more placebo than dalteparin-treated patients (25.8% vs 11.8%, P = 0.017). The second study had a similar protocol but dextran was not included; 281 patients were randomized to dalteparin or placebo and 215 were evaluable for thrombosis [37]. Significantly more placebo than dalteparin-treated patients had DVTs (11.8% vs 4.4%, P = 0.039); bleeding complications and other adverse events were similar in the two groups. The outcomes of these two studies led to the recommendation that dalteparin be continued after hospitalization. A cost-effectiveness analysis concluded that extending thromboprophylaxis for up to 4 weeks after hospital discharge was effective and cost-saving [38].

If patients given pharmacological prophylaxis during hospitalization have no evidence of VTE at the time of discharge, how likely are they to develop symptomatic thrombi? An overview analysis calculated that only 1.27% of 2361 patients with negative venography subsequently experienced clinically overt VTE [39]. Even when DVTs are detected by venography, most are non-occlusive and asymptomatic. For example, 321 asymptomatic patients had venography 10 days after hip surgery and thrombi were detected in 28% of limbs; 51% were in proximal veins and most were non-occlusive [40]. Venography might not be the best predictor of VTE risk; [41] measurements of prothrombin fragments, thrombin-antithrombin complexes, and d-dimer might be more indicative of sustained hypercoagulability; these indicators of risk were elevated in patients with thrombosis and significantly decreased by dalteparin [42]. A meta-analysis of eight randomized trials concluded that extending the duration of prophylaxis from ≤ 10 days to ≥ 21 days after hip arthroplasty significantly decreased the risk of VTE but at a cost of excess minor bleeding [43]. The decision to extend prophylaxis after hospitalization requires an assessment of the individual's risk for thrombosis and bleeding, and acceptance by a fully-informed patient.

A placebo-controlled trial of **tinzaparin**, another LMWH, found that daily doses of 50 IU/kg were associated with significantly fewer DVTs (31% vs 45%, P = 0.02), and with a non-significant increase in bleeding [44]. Tinzaparin, 75 IU/kg, was compared with adjusted-dose warfarin in a randomized, double-blind trial in 1436 patients having hip or knee arthroplasty. The incidence of DVTs was lower with tinzaparin than warfarin (31.4% vs 37.4%, P = 0.03) but the frequency of major bleeding was higher (2.8% vs 1.2%, P = 0.04) [45]. In another trial, tinzaparin, 4500 U daily, was compared with enoxaparin, 40 mg daily; the first injection was given 12 hours pre-operatively [46]. The outcomes of DVTs (21.7% and 20.1%) and bleeding were not significantly different. **Bemiparin** LMWH, 3500 IU daily, was compared with UH, 5000 IU twice daily, in a randomized, double-blind study of 300 patients and significantly decreased VTE (7.2% vs 18.7%, P = 0.01), with no difference in bleeding complications [47]. Another LMWH, **semuloparin**, was compared with enoxaparin in two trials; in SAVE-HIP1, the incidence of a composite of DVT, non-fatal pulmonary embolism, and death, as well as major bleeding was lower, but in SAVE-HIP2, the incidence of the composite endpoint was similar and clinically relevant bleeding tended to be greater [48].

The safety and efficacy of **fondaparinux** have been evaluated for the prevention of VTE after hip surgery. A double-blind, randomized trial assigned 593 patients to enoxaparin, 30 mg every 12 hours, or daily doses of fondaparinux ranging from 0.75 to 8.0 mg [49]. Fondaparinux doses ≥ 6.0 mg were associated with unacceptable bleeding; a dose of 3 mg was associated with fewer DVTs than enoxaparin (1.7% vs 9.4%, P = 0.01) with similar rates of bleeding. In follow-up studies, 2.5 mg of fondaparinux given postoperatively were compared with enoxaparin, 40 mg started pre-operatively or 30 mg given twice daily post-operatively, in patients having hip replacement [50,51]. VTE occurred in fewer patients receiving fondaparinux than enoxaparin, 40 mg (4% vs 9%) or 30 mg twice daily (6% vs 8%), with no increase in clinically relevant bleeding.

A study of patients having hip-fracture observed that fondaparinux, 2.5 mg daily, was associated with fewer DVTs than enoxaparin, 40 mg initiated pre-operatively and then given daily (8.3% vs 19.1%, $P < 0.001$), with no increase in bleeding [52]. To determine if prolonging the duration of anticoagulant prophylaxis was beneficial and safe, investigators conducted a randomized, double-blind trial in 428 patients following hip fracture surgery [53]. All had received fondaparinux, 2.5 mg, for 6−8 days and then either continued on the drug or received placebo for the next 19−23 days. Fewer patients assigned to fondaparinux developed VTE (1.4% vs 35%, $P < 0.001$), including symptomatic thrombosis (0.3% vs 2.7%, $P = 0.02$). There was a trend toward more major bleeding with fondaparinux ($P = 0.07$); this was confined to the surgical site and did not lead to reoperation, critical organ bleeding, or death. Thus, this study clearly showed the benefit of continuing thromboprophylaxis for up to one month after hip fracture surgery. A cost-effectiveness analysis showed that fondaparinux would be cost-saving as compared to enoxaparin because it might prevent an additional 30 VTE events in 1000 patients undergoing hip fracture surgery [54]. The currently recommended fondaparinux dosing for thromboprophylaxis in patients having hip surgery is 2.5 mg beginning 6−8 hours post-operatively and continued daily for up to 11 days; after hip fracture surgery, dosing is continued for up to 24 additional days.

More recently, a few **direct oral anticoagulants** have been approved for the prevention of thrombosis after hip replacement surgery. Two clinical trials randomized 6579 patients to either **rivaroxaban**, 10 mg daily beginning 6−10 hours after surgery, or enoxaparin, 40 mg started 12 hours preoperatively and continued daily [55,56]. After 13 days, the enoxaparin was discontinued and patients re-randomized to continue rivaroxaban or receive placebo until day 33, and venography was performed on day 36. The incidence of major VTE was lower with rivaroxaban than enoxaparin (0.2% vs 2.1%; 0.7% vs 4.8%; $P < 0.001$) with no significant difference in major or total bleeding. Most of the thrombotic events occurred after the discontinuation of enoxaparin on day 13. Another clinical trial randomized 3866 patients to **apixaban**, 2.5 mg twice daily, or enoxaparin, 40 mg daily; both were continued for 32−38 days [57]. Major VTE occurred in fewer patients assigned to apixaban than enoxaparin (0.5% vs 1.1%, $P = 0.01$) and with a trend toward decreased symptomatic VTE events (0.1 vs 0.4, $P = 0.11$), the percentages of patients with major bleeding were similar (0.8% vs 0.7%). **Dabigatran**, an oral thrombin inhibitor, has been compared with enoxaparin in 1777 patients undergoing hip replacement surgery [58]. Participants assigned to dabigatran received either 150 mg or 220 mg daily, starting with a half-dose 1−4 hours after surgery, or enoxaparin 40 mg beginning 12 hours before surgery; the anticoagulants were given for 28−35 days. The incidences of major VTE in the dabigatran low dose, high dose, and enoxaparin groups were 4.3%, 3.1%, 3.9%, and major bleeds were 1.3%, 2.0%, and

1.6%; the differences between groups were not statistically significant. These initial studies suggest that direct oral anticoagulants have a favorable risk/benefit profile in patients undergoing hip surgery, and further trials are anticipated.

Knee arthroplasty and knee arthroscopy

Total knee arthroplasty is complicated by a high incidence of DVT; a study that randomized 22 patients to infusions of a low-molecular-weight dextran alone recorded a prevalence of 82% [59]. The investigators observed that antithrombin levels declined after surgery; when they administered a combination of antithrombin and UH, 5000 U twice daily, the DVT rate decreased to 25%. Although most thrombi are in veins below the knee, another study showed that intraoperative tourniquet deflation is associated with showers of echogenic material visible in the right atrium and ventricle and persisting for 3−15 minutes [60]. A randomized, double-blind, placebo-controlled trial evaluated prophylaxis with **enoxaparin**, 30 mg every 12 hours started the morning after surgery and continued for up to 14 days. The risk reduction using enoxaparin was 71% (65% vs 19%, P < 0.001), and there was no increase in bleeding [61]. An open-label study randomized 453 patients to receive enoxaparin or UH, 5000 U every 8 hours [62]. The incidence of DVT was significantly lower with enoxaparin than with heparin (24.6% vs 34.2%, P = 0.02) with no difference in major bleeding. Enoxaparin was compared with warfarin in a randomized, double-blind trial; enoxaparin was significantly more effective than warfarin in preventing DVT (36.9% vs 51.7%, P = 0.003), with no difference in the incidence of clinically overt hemorrhage [63].

To ascertain whether distal thrombi detected by venography might progress after anticoagulants were withdrawn, compression ultrasonography was performed before discharge in 1984 consecutive patients having hip (1142) or knee (842) arthroplasty [64]. All patients received enoxaparin prophylaxis for 9 days and were followed for 84 days. Symptomatic venous thromboembolism (VTE) developed in 4.1%, 2.1% before and 2.0% after discharge, and only 0.15% had abnormal pre-discharge ultrasonography. Fatal pulmonary embolism was recorded in 0.15% (all after knee arthroplasty) and major hemorrhage occurred in 2.9%. The low incidence of symptomatic VTE and hemorrhage suggested that 9 days of enoxaparin is adequate and pre-discharge ultrasonography is not justified. Although the prevalence of bleeding far outweighs the risk of fatal pulmonary embolism, Hull [65] has argued that policies that emphasize bleeding avoidance at the expense of thrombosis avoidance be reconsidered. Preventing just one fatal pulmonary embolism might justify the minor bleeding risks associated with extending anticoagulant prophylaxis for a few more days. Again, this is a decision requiring individual risk factor assessment and patient input.

Other LMWHs have been evaluated for thrombus prevention in patients having knee arthroplasty. A double-blind trial randomized 246 patients to **ardeparin**, 50 U/kg twice daily, plus graduated compression stockings or compression stockings plus placebo; treatment continued for 14 days or until hospital discharge [66]. There was no difference in major bleeding, but ardeparin significantly decreased the percentage of patients with DVTs (29% vs 58%, P < 0.001). In another study, the same dose of ardeparin was compared with adjusted-dose warfarin; although blood loss was greater with ardeparin than with warfarin, the prevalence of DVT was lower (27% vs 38%, P = 0.019) [67]. Extended treatment with ardeparin failed to significantly decrease the incidence of symptomatic VTE or death [68].

Nadroparin LMWH, 60 IU/kg daily, was compared with dose-adjusted acenocoumarol in 517 patients undergoing hip or knee replacements; both agents were started preoperatively and continued for 10 days. The incidence of DVT (17% and 20%) and clinically important bleeding (1.5% and 2.3%) were not significantly different [69]. Another LMWH, **bemiparin**, was compared with enoxaparin in a randomized, double-blind trial [70]. Bemiparin, 3500 IU, was given 6 hours postoperatively, and enoxaparin, 40 mg, was started 12 hours preoperatively; both LMWHs were given daily for 10 days. There was no significant difference in the incidence of VTE or major bleeding. A meta-analysis that included all trials of LMWH for knee arthroplasty published prior to 1997 calculated a relative risk compared with adjusted dose UH or warfarin of 0.73 (95% CI, 0.66−0.80), confirming the efficacy of LMWH, and without an excess of bleeding [71]. **Semuloparin** was compared with enoxaparin in 1150 patients having total knee arthroplasty; no significant differences were observed in the composite outcome of DVT, non-fatal pulmonary embolism, or death, but clinically relevant bleeding was increased [48].

Fondaparinux, 2.5 mg daily beginning 6 hours postoperatively, was compared with enoxaparin, 30 mg twice daily, beginning 12−24 hours postoperatively in a randomized, double-blind trial encompassing more than 1000 patients having knee arthroplasty [72]. Prophylaxis was continued for up to 9 days and the duration of follow-up was 6 weeks. VTE occurred in fewer patients receiving fondaparinux than enoxaparin (12.5% vs 27.8%, P < 0.001), but major bleeding was more common (2.1% vs 0.2%, P = 0.006), although the frequency of reoperation, critical organ bleeding, or death did not differ between the two groups.

A meta-analysis of the four trials comparing fondaparinux with enoxaparin (hip replacement, knee surgery, and surgery for hip fracture) found that fondaparinux significantly reduced the incidence of VTE by day 11 (6.8% vs 13.7%, P < 0.001), but was associated with more major bleeding (P = 0.008), although the incidence of clinically relevant bleeding did not differ between groups [73]. An analysis of additional events per number of patients treated showed that the use of fondaparinux would result in one symptomatic event per 370 patients, one major bleeding event per 109 patients, and avoid one proximal thrombosis per 61 patients, and one VTE per 14 patients [74]. This review supports the conclusion that greater protection against thrombosis comes at the expense of more bleeding.

As with LMWHs, the issues addressed regarding fondaparinux are the timing of the initial dose, the duration of treatment, and its cost-effectiveness. An open-label trial randomized 2000 hip and knee replacement patients to receive fondaparinux 6−8 hours after surgery or the following morning; there were no significant differences in thrombosis and a trend toward fewer clinically relevant hemorrhages in the delayed treatment group [75]. An analysis of the four trials in the 7344 patients described previously observed that the efficacy of fondaparinux and enoxaparin was significantly increased by longer treatment duration; increasing the duration of therapy from ≤5 days to 9−11 days decreased the VTE incidence from 8.7% to 5.2% with fondaparinux, and from 17% to 11.7% with enoxaparin [76]. In patients receiving 3−5 weeks of fondaparinux following hip fracture surgery, temporary discontinuation of the anticoagulant for 48 hours permitted safe removal of neuraxial or deep peripheral nerve catheters without decreasing efficacy or inducing a perineural hematoma [77]. Lastly, fondaparinux was found to be cost-saving because it was associated with fewer clinical VTE events than enoxaparin [78].

Detractors of fondaparinux have argued that by combining symptomatic with asymptomatic VTE, the clinical trials masked the fact that symptomatic events were more frequent in the fondaparinux group [79]. Another concern is that bleeding with fondaparinux might be exacerbated by the long half-life of the drug and the lack of a reversal agent. Additionally, fondaparinux is more expensive than LMWHs. Nevertheless, the simplicity of administration (2.5 mg given subcutaneously once daily) and general effectiveness have established the place of fondaparinux in thromboprophylaxis.

Thromboprophylaxis with **direct oral anticoagulants** has been studied in patients undergoing total knee replacement. A double-blind trial randomized 1684 patients to **rivaroxaban**, 10 mg daily beginning 6−10 hours after surgery, or enoxaparin, 40 mg started 12 hours preoperatively; both were continued daily for 12 days [80]. Major VTE and symptomatic VTE were less common with rivaroxaban than enoxaparin (1.0% vs 2.5%, P = 0.02% and 0.7% vs 2.0%, P < 0.005, respectively), and major bleeding was <1% in each group. Two double-blind trials randomized 4260 patients undergoing knee replacement surgery to **apixaban**, 2.5 mg twice daily, or enoxaparin, 30 mg every 12 hours in one trial and 40 mg daily in the second study [81,82]. In the first study, the percentages of patients in the apixaban and enoxaparin groups with major VTE were similar but major hemorrhages occurred less often with apixaban (0.7% vs 1.4%, P = 0.05), while in the second study, which used a smaller dose of enoxaparin, major VTE was significantly decreased with apixaban (1.1% vs 2.17%, P = 0.02), while major bleeding events did not differ. Another double-blind study randomized 2076 patients to **dabigatran**, 150 mg or 220 mg daily starting with a half-dose 1−4 hours after surgery, or enoxaparin, 40 mg daily beginning the evening before surgery [83]. In the dabigatran groups (150 mg or 220 mg) and enoxaparin group, major VTE occurred in 3.8%, 2.6%, and 3.5%, and major bleeding in 1.3%, 1.5%, and 1.3%; the differences between groups were not significant.

Systematic reviews and meta-analyses have compared the effectiveness of direct oral anticoagulants and LMWH in the prevention of VTE after hip and knee replacement. A study published in 2012 concluded that the use of a factor Xa inhibitor in 1000 patients would result in four fewer symptomatic DVT events at a cost of 2 major bleeding events [84]. A subsequent review noted that the VTE risk was non-significantly reduced by rivaroxaban compared with dabigatran and apixaban, but with increased bleeding [85]. More recently, a network meta-analysis reported that direct oral factor Xa inhibitors had a more favorable hemostatic profile than LMWHs, preventing 4-fold more symptomatic DVTS without significantly more bleeding [86]. From an economic perspective, the modestly increased effectiveness of factor Xa inhibitors translated into slightly lower average costs per patient [87].

Knee arthroscopy is usually performed in conjunction with meniscectomy, synovectomy, or reconstruction of the cruciate ligaments. Graduated compression stockings and early mobilization are recommended for thromboprophylaxis, but whether anticoagulants might confer additional benefit is uncertain. This issue was addressed by a controlled clinical trial that randomized 1761 patients to compression hose or nadroparin LMWH, 3800 IU daily for 7 days [88]. The 3-month cumulative incidence of the primary endpoint, a composite of asymptomatic proximal DVT, symptomatic VTE, and death, was higher in the stockings group than the nadroparin group (3.2% vs 0.9%, P = 0.005), and major bleeding rates were similar. However, there were only eight thrombi in proximal veins (1.2% of the total DVT), and seven of the eight were asymptomatic [89]. A large open-label trial randomized almost 3000 patients to no anticoagulant or a LMWH (nadroparin, 2850 IU or dalteparin, 2500 IU, each given daily for 8 days), and found no difference in VTE or bleeding [90]. Although guidelines recommend against the use of anticoagulants in patients without a history of VTE [91], the decision should be made on an individual basis that includes the patient's propensity for thrombosis and bleeding [92]. In addition, the risks of thrombosis associated with the procedure as well as the potential for hemorrhage with anticoagulants should be discussed, and patient preferences respected [93].

Lower extremity DVT is a frequent complication of major trauma, with proximal thrombi reported in 18% and total DVT in 58% [94]. A double-blind trial randomized 344 trauma patients to enoxaparin, 30 mg every 12 hours, or UH, 5000 IU every 12 hours, beginning 36 hours after injury (intracranial bleeding was an exclusion) [95]. Venography by day 14 revealed that fewer patients given enoxaparin had DVT (31% vs 44%, P = 0.014), with a trend toward more major bleeding in the enoxaparin group (2.9% vs 0.6%, P = 0.12). Thrombus prevention in patients with lower extremity immobilization has also been examined. An open-label trial in 339 patients with plaster cast immobilization of the legs reported no symptomatic DVTs in patients randomized to LMWH compared to 4.3% of controls (P < 0.007), but the duration of casting was shorter in the former (15.2 days vs 18.8 days, P < 0.01) [96]. Subsequently, a prospective, double-blind trial randomized 371 patients with leg injuries requiring immobilization to a LMWH (reviparin) or placebo [97]. DVT occurred in fewer patients assigned to reviparin than placebo (9% vs 19%, P = 0.01); the frequency of bleeding or other adverse events was similar in the two groups. The efficacy of LMWH was supported by a meta-analysis of six trials that randomized 1456 patients; the incidence of asymptomatic VTE was significantly lower with LMWH than with placebo (17.1% vs 9.6%, P = 0.006), with no difference in bleeding [98]. A Cochrane Database Systematic Review also concluded that LMWH reduced the incidence of DVT in outpatients with immobilized lower limbs [99]. **Fondaparinux** was compared with **nadroparin** in an open-label study of 1170 patients with rigid or semi-rigid immobilization for non-surgical below-knee injuries [100]. Fewer VTE occurred in those assigned to fondaparinux than nadroparin (2.6% vs 8.2%, P < 0.001), with only one major hemorrhage. In summary, these studies confirm that LMWH or fondaparinux safely reduces the risk of VTE in immobilized patients with major trauma who do not have intracranial bleeding.

Abdominal surgery

This category includes abdominal, gynecologic, or urologic operations lasting more than one hour. Thromboprophylaxis with UH, 5000 IU subcutaneously every 12 hours, is generally safe and effective in patients without a bleeding tendency [101]. Large, randomized trials have been conducted to determine whether LMWH is more advantageous than UH. A study by Kakkar et al. [102] randomized 3809 patients to **dalteparin**, 2500 IU daily, or UH, 5000 IU twice daily, with the first dose of each given 1—4 hours preoperatively. VTE incidence was about 1% in each group but severe bleeding and wound hematoma were significantly less frequent in the dalteparin group. A double-blind study that compared dalteparin doses of 2500 IU and 5000 IU, given the evening prior to surgery and then daily, reported a significantly lower DVT incidence with the 5000 IU dose (6.8% vs 13.1%, P < 0.001) but higher bleeding rate (4.7% vs 2.7%, P = 0.02) [103]. Another study showed that continuing dalteparin, 5000 IU daily, for 28 days as compared to 7 days, decreased the cumulative incidence of VTE in the long-term group (7.3% vs 16.3%, P = 0.012), without increasing bleeding events [104]. The currently recommended doses of dalteparin are 2500 IU or 5000 IU once daily, or 2500 IU followed by 2500 IU 12 hours later and then 5000 IU once daily.

Enoxaparin has also been compared with UH. A multicenter controlled trial randomized 1122 patients to enoxaparin, 20 mg daily, or UH, 5000 IU twice daily, beginning 2 hours preoperatively and continued daily [105]. VTE was infrequent with either regimen, but there was less pain and bleeding at the enoxaparin injection site. A double-blind study randomized 718 patients to enoxaparin, 20 mg daily, and 709 patients to UH, 5000 IU every 8 hours [106]. The incidence of DVT was similar in both groups (8.1% and 6.3%), as was major bleeding (1.5% and 2.5%), but re-operation for bleeding occurred less often with enoxaparin than with UH (0.6% vs 1.8%, P = 0.03). Enoxaparin, 40 mg daily, was compared with UH, 5000 IU every 8 hours, in a double-blind trial encompassing 936 patients undergoing colorectal surgery [107]. No differences in the rates of VTE, major bleeding, or re-operation for bleeding were observed. A trial that compared enoxaparin, 40 mg daily started preoperatively, with **semuloparin**, 20 mg daily started postoperatively, reported similar DVT rates but less major bleeding with semuloparin (2.9% vs 4.5%) [108]. Current recommendations for dosing enoxaparin are 40 mg once daily initiated two hours preoperatively and continued for 7−10 days.

Fondaparinux combined with intermittent pneumatic compression (IPC) was compared with intermittent pneumatic compression (IPC) alone in a randomized double-blind study of 842 patients undergoing abdominal surgery [109]. The use of fondaparinux significantly decreased the VTE rate (1.7% vs 5.3%, P = 0.004) and proximal DVT (0.2% vs 1.7%, P = 0.037) but increased major bleeding (1.6% vs 0.2%, P = 0.006), although none of the hemorrhages were fatal or in a critical organ. A randomized, double-blind trial compared fondaparinux, 2.5 mg daily beginning 6 hours postoperatively, with dalteparin, 2500 IU given 2 hours before and 12 hours after surgery and then 5000 IU daily; treatment continued for 5−9 days [110]. The VTE rates (4.6% vs 6.1%) and major bleeding events (3.4% and 2.4%) did not differ significantly between fondaparinux and dalteparin. The recommended dose of fondaparinux is 2.5 mg daily beginning 6−8 hours postoperatively and continued for 5−9 days.

The risk for postoperative VTE is increased with cancer surgery, especially when major operations such as radical mastectomy or prostatectomy are performed. The ENOXACAN Study Group conducted a double-blind trial that randomized 631 patients undergoing curative cancer surgery to **enoxaparin**, 40 mg daily, or UH, 5000 IU three times daily, each beginning two hours before surgery and continuing until venography was performed on day 10 [111]. The incidences of DVTs and major bleeding with enoxaparin and UH were similar (14.7% and 18.2%; 4.1% and 2.9%), and there were no differences in mortality. A second trial by the same investigators examined whether extending enoxaparin prophylaxis for an additional 21 days would be beneficial [112]. Enoxaparin was given for 6−10 days postoperatively and then 501 patients were randomized to continue enoxaparin or switch to placebo, and venography was performed between days 25 and 31. DVT was less frequent in the enoxaparin group (4.8% vs 12%, P = 0.02), and bleeding was not significantly increased. Using a similar protocol, 625 cancer patients undergoing abdominal or pelvic surgery were randomized to **bemiparin**, 3500 IU daily, or placebo [113]. Major VTE was decreased in the bemiparin group (0.8% vs 4.6%, P = 0.01) and major bleeding was not increased.

A double-blind trial randomized 950 patients undergoing resection for colorectal cancer to **nadroparin**, 2850 IU, or enoxaparin, 40 mg, daily for 9 days [114]. The overall VTE rate was higher with nadroparin (15.9% vs 12.6%), but there were fewer symptomatic events (0.2% vs 1.4%) and significantly less major bleeding (7.3% vs 11.5%, P = 0.01). Other LMWHs such as **tinzaparin** and **certoparin** also have been found to be efficacious in abdominal surgery. A meta-analysis concluded that LMWH was as safe and effective as UH in preventing DVT after surgery, but associated with fewer pulmonary emboli and injection site hematomas [115]. A subsequent analysis of 14 clinical trials observed that mortality was similar with various doses of LMWH or UH, but there were fewer DVTs with LMWH than UH (3.9% vs 6.1%, P = 0.02), and no difference in the frequencies of pulmonary embolism and bleeding [116].

Patients with cancer comprised 68% of those undergoing abdominal surgery in the trial described previously, and those assigned to **fondaparinux** had a lower postoperative VTE rate than those receiving enoxaparin (4.7% vs 7.7%), with no difference in major bleeding [110]. A study of surgery for urologic malignancies reported similar rates of bleeding with fondaparinux, 2.5 mg daily, and enoxaparin, 20 mg twice daily [117]. A randomized double-blind study of patients having minimally invasive esophagectomy compared the safety and efficacy of daily doses of fondaparinux and nadroparin, 2850 IU [118]. Ultrasound evaluation on day 7 detected DVTs in fewer patients receiving fondaparinux (1.69% vs 12.28%, P = 0.03), and neither group had bleeding events.

A Cochrane Database Systemic Review included 20 randomized clinical trials encompassing 9771 patients with cancer receiving perioperative thromboprophylaxis [119]. LMWH was compared with UH and fondaparinux, and no differences were observed in their effects on mortality, VTE, major or minor bleeding, but wound hematomas were less frequent with LMWH than with UH. Anticoagulant prophylaxis is indicated for inpatients scheduled for major operations (procedures >30 minutes duration, usually under general anesthesia) who are at moderate to high risk for VTE (older, obese, presence of malignancy, previous thrombi); those at highest risk should continue thromboprophylaxis for an extended 4-week period [120].

Cardiac surgery

Coronary artery bypass graft placement is a common surgical procedure that is infrequently complicated by postoperative VTE. A comparative effectiveness study found that chemical or mechanical preventive measures did not appreciably reduce the already very low incidence of VTE [121]. DVTs and pulmonary emboli occur more often with other cardiopulmonary operations, so prophylaxis with heparins is often prescribed. An open-label study randomized 2551 patients undergoing open-heart surgery to UH, 5000 IU every 12 hours, with or without pneumatic compression stockings [122]. Symptomatic pulmonary embolism occurred less frequently in the group wearing compression stockings (1.5% vs 4%, P < 0.01). Therefore, because cardiac surgery patients are at a relatively high risk of postoperative bleeding, the American College of Chest Physicians recommended only mechanical prophylaxis for patients with an uncomplicated postoperative course, and the addition of UH or LMWH if the hospital course was prolonged by one or more non-hemorrhagic surgical complications [123]. But a more recent systematic review concluded that pharmacologic prophylaxis significantly reduced the risk of pulmonary embolization and symptomatic VTE, especially in the presence of risk factors such as previous VTE, obesity, heart failure, immobilization, and mechanical ventilation, with no increase in bleeding attributable to the prophylactic agents [124].

Neurosurgery

Brain tissue has long been recognized as being extremely thrombogenic; fragments entering the blood stream as a consequence of surgery or trauma might provoke thrombosis. This risk for VTE is enhanced by paralysis or immobilization of the lower extremities, warranting consideration for thromboprophylaxis. Although anticoagulants are generally more effective than compression devices or hose, the possibility of intracranial or intraspinal hemorrhage has limited their application. To mitigate this risk, a trial conducted in 1978 screened potential participants with a battery of clotting assays and a test for heparin sensitivity [125]. They then randomized 100 low risk patients undergoing elective neurosurgery to placebo or UH, 5000 IU two hours preoperatively and every 8 hours postoperatively, continued for 7 days. The incidence of DVT was lower in the UH group (6% vs 34%, P < 0.005), and there were no differences between groups in postoperative hemoglobin, transfusion requirements, or wound hematomas.

A study in 307 patients undergoing elective neurosurgery showed that compression hose and **enoxaparin**, 40 mg daily beginning 24 hours after surgery, significantly decreased both total DVT and proximal venous thrombosis as compared to compression hose alone (17% vs 32%, P = 0.004; and 5% vs 13%, P = 0.04), with similar rates of bleeding [126]. However, if enoxaparin was given prior to the induction of anesthesia, the incidence of clinically significant intracranial hemorrhage was increased [127]. **Dalteparin**, 2500 IU daily, was compared with UH, 5000 IU every 12 hours, in a randomized trial of 100 patients undergoing craniotomy [128]. Both anticoagulants were begun at the initiation of anesthesia and continued for 7 days; patients with hypersensitivity to heparin were excluded and all patients wore IPC devices. VTE and major hemorrhages were infrequent and not significantly different in the two groups. A double-blind study randomized 485 patients undergoing craniotomy or spinal column surgery to compression hose alone or compression hose with **nadroparin**, 7500 IU started 18−24 hours postoperatively and then daily until discharge [129]. The main exclusion criteria were an underlying bleeding tendency or the presence of an aneurysm. VTE incidence was lower in the nadroparin group (18.7% vs 26.3%, P = 0.047), and major bleeding was not significantly different. A meta-analysis of 30 studies comprising 7779 patients showed that LMWH and intermittent compression devices were effective in preventing DVTs, and the rate of intracranial hemorrhage (ICH) did not differ significantly between the two modalities, although the pooled rate of ICH and minor bleeding was higher with heparins [130].

A 2012 analysis of trial data concluded that LMWH might prevent between 8 and 36 VTE events at a cost of 4−22 intracranial hemorrhages and led to the recommendation for the use of only IPC in patients undergoing craniotomy [123]. However, it was suggested that anticoagulants be added to IPC for patients at very high risk of VTE, although a trial that randomized patients with malignant glioma (a very high risk group) to dalteparin (5000 IU daily) or placebo started within 4 weeks of surgery reported that the decreases in VTE and increases in intracranial bleeding with dalteparin were not statistically significant (P = 0.29) [131]. In summary, all patients should be fitted with IPC devices, and if the risk of VTE is high, consider starting LMWH no sooner than 24 hours postoperatively. Anticoagulants are contraindicated in patients with unclipped or multiple cerebral aneurysms, necrotic tissue, or vascular metastasis.

Acute spinal cord injury

Individuals with lower extremity paralysis due to acute spinal cord injury usually require spine stabilization surgery. Anticoagulants initiated within 72 hours of injury mitigate the exceptionally high risk of VTE in these patients. An open-label trial randomized 107 patients to **enoxaparin**, 30 mg every 12 hours, or UH, 5000 IU every 8 hours for 14 days; all patients wore IPC devices [132]. Although the overall incidence of VTE was similar in the two groups, fewer patients assigned to enoxaparin had pulmonary emboli (5.2% vs 18.4%, P − 0.03) and the frequency of major bleeding was not significantly increased. Prophylaxis was continued for six additional weeks; the incidence of new VTE trended to be lower in those receiving enoxaparin as compared with heparin (8.5% vs 21.7%, P = 0.05) with no difference in bleeding events [133]. Another open-label study randomized 95 patients to dalteparin, 5000 IU daily, or enoxaparin, 30 mg every 12 hours, for up to 3 months [134]. The incidences of DVT (4% and 6%) were low as was bleeding (4% and 2%). The use of combined mechanical (IPC) and anticoagulant (UH, LMWH) prophylaxis for these high risk patients is recommended [121].

Medical inpatients

Individuals >40 years of age, hospitalized and bedridden because of congestive heart failure, respiratory failure, infections, rheumatologic disorders, or inflammatory bowel disease, are vulnerable to thromboembolic events. Bilateral venography or duplex ultrasound performed after a week of hospitalization detects DVTs in nearly 15% of these patients [135]. Fatal pulmonary embolism occurs in 0.48% of elderly patients hospitalized with infections [136]. To examine the efficacy of prophylaxis, inpatients with even-numbered hospital records were given UH, 5000 U twice daily and those with odd-numbered records were untreated [137]. Mortality was decreased in the low-dose heparin group (7.8% vs 10.9%; P = 0.025). Subsequent studies of UH did not find a long-term (> 30 days) reduction of mortality, although most of the fatal pulmonary emboli occurred after UH was discontinued at day 21 [136].

To assess the efficacy of thromboprophylaxis after acute myocardial infarction, a double-blind study randomized 146 patients to placebo or UH, 5000 IU every 12 hours for 10 days [138]. DVTs were decreased in the heparin group (3.2% vs 17.2%, P < 0.025), and there was no major bleeding. Another double-blind study randomized 866 patients with acute medical illnesses to placebo or **enoxaparin**, 20 mg or 40 mg daily, and reported that VTE was significantly reduced in the enoxaparin 40 mg group compared to placebo (5.5% vs 14.9%, P < 0.001) [126]. However, mortality at 110 days was similar in the three groups, a result that was confirmed by a subsequent double-blind trial that compared a 30- day course of enoxaparin, 40 mg daily, with placebo; the death rates were 4.9% and 4.8% [139]. A meta-analysis of four trials that compared enoxaparin with UH observed risk reductions of 37% for all VTE and 62% for symptomatic VTE with enoxaparin, with low major bleeding rates [140]. To assess the efficacy and safety of a 4-week extension of enoxaparin beyond an initial 10 day course, nearly 6000 patients were enrolled in a double-blind, placebo controlled trial [141]. VTE and symptomatic DVT were significantly decreased (P < 0.05) with enoxaparin compared to placebo (2.5% vs 4% and 0.2% vs 1.2%, respectively), but there were more major hemorrhages (0.8% vs 0.3%, P < 0.05). Subgroup analyses showed that the benefit of extended duration therapy was confined to women, patients >75 years of age, and those most immobile, albeit at a small increased risk of bleeding [142]. Enoxaparin, 40 mg daily for 6−11 days, is the dosing recommended by the manufacturer for medical patients with restricted mobility during acute illness.

A double-blind trial randomized 3706 patients with acute medical illnesses to placebo or **dalteparin**, 5000 IU daily for 14 days [143]. VTE was significantly decreased by dalteparin (2.77% vs 4.96%, P = 0.001), but there was a trend toward more major bleeding (0.49% vs 0.16%, P = 0.15). Dalteparin, 5000 IU daily, was compared with UH, 5000 IU twice daily, in a randomized trial of more than 3700 patients in intensive care units [144]. Fewer patients in the dalteparin group than in the UH group experienced pulmonary emboli (1.3% vs 2.3%, P = 0.01), although there was no difference in the incidence of proximal vein thrombosis (∼5% in each group). The between-group differences in death and major bleeding were similar, but fewer patients in the dalteparin group developed heparin-induced thrombocytopenia (0.3% vs 0.6%). A subgroup analysis of patients with endstage renal disease or severe renal insufficiency found no difference between dalteparin and UH in the incidence of VTE or major bleeding, although dalteparin-treated patients with renal dysfunction had more proximal DVTs than those receiving UH [145]. Dalteparin, 5000 IU daily, is the manufacturer's recommended dose for thrombosis prevention in medical patients.

Fondaparinux has been evaluated in older patients with acute medical illnesses. A placebo-controlled trial randomized 849 patients to placebo or fondaparinux, 2.5 mg daily for 6−14 days [146]. VTE incidence was reduced in the fondaparinux group (5.6% vs 10.5%, P = 0.029), and the fatal pulmonary emboli (5) were confined to the placebo group. There was no difference in the incidence of major bleeding. A systematic review and meta-analysis of trials published in 2008 reported that anticoagulant prophylaxis conferred an absolute risk reduction of 2.6% for any DVT and 1.8% for proximal DVT, but was associated with a 0.5% increased risk of major bleeding [147].

Direct oral anticoagulants have been compared with LMWH. A double-blind trial randomized 4495 patients to **apixaban**, 2.5 mg twice daily for 30 days, or enoxaparin, 40 mg daily for 6−14 days [148]. VTE were detected in similar percentages of patients in the two groups, but major bleeding by day 30 was more common with apixaban (0.47% vs 0.19%, P = 0.04). Another clinical trial of 8101 patients aged ≥ 40 reported that the composite endpoint of symptomatic VTE or asymptomatic proximal DVT occurred with equal frequency (2.7%) in patients receiving **rivaroxaban**, 10 mg daily, or enoxaparin, 40 mg daily, during the first 10 days following randomization, but a composite of major or clinically relevant nonmajor bleeding was more frequent in the rivaroxaban group (2.8% vs 1.2%, P < 0.001) [149]. During the next 25 days, patients in the enoxaparin group were switched to placebo and had more thrombotic episodes than patients continued on rivaroxaban (5.7% vs 4.4%, P = 0.02) but fewer bleeding events (1.7% vs 4.1%, P < 0.001).

Post-hospitalization thromboprophylaxis with rivaroxaban was evaluated in 12,019 patients at increased risk for VTE [150]. They were randomly assigned to rivaroxaban, 10 mg daily, or placebo in a blinded, 45 day trial. The primary efficacy outcome of symptomatic VTE or death was similar in the two groups, but symptomatic nonfatal VTE occurred less often with rivaroxaban (0.18% vs 0.42%, hazard ratio 0.44), and major bleeding was infrequent (0.28%). Another clinical trial with a similar design compared **betrixaban**, 80 mg daily, with enoxaparin in 7513 acutely ill medical patients [151]. Symptomatic VTE or asymptomatic proximal DVT was significantly less frequent in patients aged ≥ 75 years, with no increase in major bleeding. Subgroup analyzes showed that betrixaban reduced the incidence of fatal or irreversible acute events, and in patients with prior VTE, the frequency of recurrences [152,153]. A systematic review and meta-analysis of four trials that included 34,068 medical patients reported that extended-duration treatment decreased the risks of symptomatic DVT and non-fatal pulmonary embolism. The risk of major bleeding was increased with apixaban, enoxaparin, and rivaroxaban, but not with betrixaban [154].

Current guidelines from the American Society of Hematology conditionally recommend that acutely ill medical patients, and especially those with critical illnesses, receive prophylaxis with LMWH or fondaparinux rather than UH or direct oral anticoagulants [155]. If the risk of bleeding is unacceptable, pneumatic compression devices or graduated compression stockings are recommended. Although extended-duration outpatient treatment with direct oral anticoagulants was not recommended, the panel might reconsider this guideline given the outcome of the betrixaban trial.

Cancer inpatients and outpatients

The VTE risk in patients with cancer is increased and related to the type and location of the tumor; patient age and body mass index; treatment modality (chemotherapy, radiation, surgery); and other factors. For example, patients receiving chemotherapy are at increased risk if they have anemia, leukocytosis and thrombocytosis [156]. Patients with multiple myeloma, most of whom receive immunomodulatory agents, have a 9-fold increased risk of DVT [157]. Several clinical trials have evaluated the safety and efficacy of heparins in patients with malignancies.

An open-label trial that randomized cancer patients to **enoxaparin** (1.5 mg/kg daily) or adjusted-dose warfarin observed a trend toward fewer hemorrhages and deaths with enoxaparin [158]. Another trial in 312 outpatients with advanced pancreatic cancer randomized patients to enoxaparin, 1.0 mg/kg daily for 3 months, or no anticoagulant [159]. There were fewer symptomatic VTEs in the enoxaparin group (1.3% vs 9.9%, P = 0.001), with no difference in major bleeding events. An open-label study randomized 98 ambulatory cancer patients at high-risk of VTE to **dalteparin**, 5000 IU daily, or observation for 12 weeks [160]. VTE trended to be less frequent with dal-teparin (12% vs 21%,) but clinically relevant bleeding was significantly increased (hazard ratio = 7). **Nadroparin**, 3800 IU daily for up to 4 months, was studied in a placebo-controlled trial of 1150 ambulatory patients receiv-ing chemotherapy [161]. VTE was less frequent in the nadroparin group (2% vs 3.9%, P = 0.02), with a low incidence of major bleeding (0.7%). However, a randomized, open-label study could not confirm that 6 weeks therapy with nadroparin provided a survival benefit in patients with advanced prostate, lung, or pancreatic cancer [162]. Furthermore, a meta-analysis of 9 trials encompassing 5987 patients found that LMWH had no effect on the survival of cancer patients [163].

Another LMWH, **semuloparin**, was evaluated in a double-blind, placebo-controlled trial in 3212 patients receiving chemotherapy for meta-static or locally advanced solid tumors [164]. Semuloparin was given in a dose of 20 mg daily for a median of 3.5 months. The incidence of VTE was decreased by semuloparin (1.2% vs 3.4%, P < 0.001); clinically relevant bleeding (2.8% vs 2%) and major bleeding (1.2% vs 1.1%) were not signifi-cantly increased. These data were reviewed by an FDA advisory committee and approval was not recommended because the 2.2% reduction in VTE was considered modest, the extent of bleeding in the placebo group was ques-tioned, and the relatively short duration of follow-up felt to be inadequate.

The issue of VTE prevention in cancer outpatients was recently addressed by two trials with direct oral anticoagulants. The first randomized 841 indivi-duals with elevated thrombosis risk scores to placebo or **rivaroxaban**, 10 mg daily, for 180 days [165]. Although similar percentages of patients in the two groups had VTE (6% and 8.8%), fewer rivaroxaban than placebo-treated patients had thrombi during cancer interventions (2.6% vs 6.4%, hazard ratio, 0.4). Differences between groups in major bleeding, clinically relevant bleed-ing, and adverse events were not significant. The second study randomized 563 intermediate to high-risk outpatients who were initiating chemotherapy to placebo or **apixaban**, 2.5 mg twice daily, for 180 days [166]. VTE occurred in fewer patients receiving apixaban than placebo (4.2% vs 10.2%, P < 0.001), but the apixaban group had more major bleeding (3.5% vs 1.8%, P = 0.046).

The efficacy of LMWHs versus UH was evaluated in critically ill patients with cancer using a propensity-matched comparative-effectiveness cohort from the Premier Database [167]. Twice as many patients were receiving LMWHs as UH (42,000 vs 21,000), and enoxaparin was prescribed more often (94%) than dalteparin (3%) or fondaparinux (3%). Although LMWHs were not associated with a decreased incidence of VTE as compared with UH, they did reduce the frequency of pulmonary embolism (0.7% vs 0.99%), significant bleeding (13.3% vs 14.8%), and heparin-induced thrombocytope-nia (0.06% vs 0.19%); all P-values <0.001.

Guidelines from the International Society of Haematology recommend that hospitalized cancer patients with acute medical illnesses receive LMWH rather than UH or direct oral anticoagulants [168]. Pneumatic compression devices are suggested for those with a high risk of bleeding, but anticoagulants should be started once the bleeding risk has resolved. Most guidelines for the use of thromboprophylaxis recommend against their routine use in ambulatory cancer patients, but they might be considered for patients with pancreatic or gastric malignancies, or non-cancer risk factors [169]. Another exception is multiple myeloma; anticoagulants (LMWH or warfarin) are recommended for patients with ≥ 2 individual or myeloma-related risk factors and/or undergoing therapy with high-dose dexamethasone, doxorubicin, or multi-agent chemotherapy [170]. Bleeding risks should be assessed and the risk-benefit ratio discussed with the patient.

Thromboprophylaxis in stroke patients

Paralytic stroke is associated with a high incidence of VTE warranting consideration of anti-thrombotic therapy. A randomized trial of low-dose UH, 5000 IU every eight hours, or no treatment was conducted in 32 elderly patients with ischemic strokes; DVTs were detected using the labeled fibrinogen test [171]. Fewer heparin-treated patients than controls had DVTs (12.5% vs 75%, P < 0.01). A double-blind trial compared UH, 5000 IU every 8 hours, with **enoxaparin**, 40 mg daily, in 212 patients who underwent bilateral venography at final assessment [172]. VTE occurred less frequently in the enoxaparin group (19.7% vs 34.7%, P = 0.04). One enoxaparin-treated patient had an intracerebral hemorrhage, but hemorrhagic transformation of the cerebral infarct was less frequent with enoxaparin than UH (13.2% vs 18.9%). A subsequent open-label study randomized 1762 patients stratified by stroke severity score to enoxaparin, 40 mg daily, or UH, 5000 IU every 12 hours for 10 days [173]. Enoxaparin use was associated with a decrease in VTE regardless of the stroke severity (10% vs 18%, P < 0.001), and the rate of symptomatic intracranial bleeding was low (≤1%) in both groups, although there were more major extracranial hemorrhages in the enoxaparin group (1% vs 0, P = 0.015). The EXCLAIM study administered enoxaparin, 40 mg daily, to hospitalized patients with acute medical illnesses for 10 days and then randomized them to enoxaparin or placebo for 28 days; 389 of the participants had ischemic strokes [174]. VTE incidence was lower in the enoxaparin group (2.4% vs 8%, P = 0.02) but was associated with an increase in major bleeding (1.5% vs 0).

A double-blind study randomized 60 patients with ischemic stroke to **dalteparin**, 2500 IU twice daily, or placebo. DVT was decreased in the dalteparin group (22% vs 50%, P = 0.05), but total bleeding was increased (23% vs 7%) [175]. Another study randomized 103 patients to dalteparin, 55−65 IU/kg (∼4200 IU) daily, or placebo until hospital discharge [176].

No differences in the incidence of DVT (36% and 34%) or major bleeding was observed. **Certoparin**, 3000 IU daily, was compared with UH, 5000 IU every 8 hours, in a double-blind, randomized trial of 545 patients with acute ischemic stroke [177]. Major bleeding as well as the primary endpoint, a composite of proximal vein thrombosis, pulmonary embolism, or VTE death, occurred with similar frequency in the two groups. A meta-analysis of three trials (references 172, 173, and 177) concluded that LMWH significantly reduced VTE risk as compared to UH without increasing bleeding [178], and a similar conclusion was reached by an analysis that included only the trials with enoxaparin [140]. Subsequently, the data on hospitalized medical patients and those with acute stroke were combined (18 studies, 36,122 patients) and analyzed [179]. Although UH prophylaxis reduced the incidence of pulmonary emboli, it increased major bleeding events. Nevertheless, there was a trend toward a reduction in total mortality ($P = 0.056$). Analysis of the 14 trials that compared UH with LMWH observed no significant differences in outcomes and formed the basis for the American College of Physicians Clinical Practice Guideline that heparin or a related drug be used for VTE prophylaxis in medical (including stroke) patients unless the bleeding risks outweigh the benefits [180].

In addition to preventing VTE, the administration of enoxaparin, 20 mg daily for three weeks, decreased cerebral vasospasm and ischemia following subarachnoid hemorrhage [181]. Furthermore, the oral anti-Xa inhibitor, betrixaban, 80 mg daily for 35−42 days, reduced the incidence of new strokes for up to 77 days during and after hospitalization for an acute medical illness [182].

Arterial occlusion

Thrombi occasionally form on proliferating vascular intima and occlude arteries; the anti-proliferative as well as anticoagulant activity of heparins might preserve the patency of these diseased vessels. In 1973, Sharnoff [183] suggested that perioperative heparin administration could prevent postoperative coronary artery thrombosis. This possibility was examined by a trial that examined whether UH might prevent recurrent myocardial infarction; 728 patients were randomized to usual treatment or UH, 12,500 IU daily for 265 days [184]. The re-infarction rate was significantly lower in the heparin group (1.32% vs 3.56%, P < 0.05), and cumulative mortality was lower (3.3% vs 6.3%, P < 0.05). No major hemorrhages or other adverse effects requiring discontinuation of therapy were reported.

Heparins have also been used to maintain vascular graft patency in patients undergoing femoropopliteal bypass grafting. In an open-label study, 200 patients were given postoperative prophylaxis with **dalteparin**, 2500 IU daily two hours preoperatively and then daily for 7 days [185]. They were then randomized to continue dalteparin or receive aspirin/dipyridamole, and assessed at 3-month intervals for 1 year. Graft survival was better in the dalteparin group (78% vs 64%); the benefit was confined to those having salvage surgery. No major bleeding events occurred in either group. Another trial randomized 201 patients to an intravenous bolus of **enoxaparin**, 75 IU/kg immediately prior to arterial cross-clamping, followed by 75 IU/kg every 12 hours for 10 days, or a bolus of UH, 50 IU/kg, followed by 150 IU/kg every 12 hours for 10 days [186]. Graft thrombosis was less frequent with enoxaparin (8% vs 22%, P = 0.009), with no difference in the frequency of major hemorrhages (12 in each group). Efforts to duplicate the benefits of systemic LMWHs with heparin-bonded polytetrafluoroethylene grafts have been only partially successful; patency was improved at 2 years, but after 5 years of follow-up was no better than that achieved with standard grafts [187].

Studies have assessed whether heparins prevent left ventricular (LV) thrombus formation after acute anterior myocardial infarction. A double-blind trial randomized 221 patients to high doses of UH, 12,500 IU, or low doses, 5000 IU, each given every 12 hours for 10 days [188]. LV thrombi were less common in the high dose group (11% vs 32%, P = 0.0004); the frequency of hemorrhagic complications was similar in the two groups. **Enoxaparin**, 1 mg/kg every 12 hours for 30 days, was compared with heparin/warfarin in a randomized trial of 60 patients with ejection fractions of ≤ 40% [189]. There were more mural thrombi in the enoxaparin group, although the difference was not statistically significant, and bleeding and thromboembolic events were rare in both groups. Another double-blind trial randomized 776 patients to **dalteparin**, 150 IU/kg every 12 hours, or placebo after thrombolytic therapy and aspirin, and echocardiography was performed on day 9 [190]. Left ventricular thrombus or embolism occurred less often in the dalteparin group (14.2% vs 21.9%, P = 0.03), but with a significant increase in major as well as minor hemorrhages (2.9% vs 0.3%, and 14.8% vs 1.8%, respectively). A recent systematic literature review concluded that the evidence supporting the use of prophylactic anticoagulation to prevent LV thrombus formation and systemic embolism was inconsistent and would require an appropriately powered randomized trial [191]. At present, either warfarin or direct oral anticoagulants are selected for this indication [192].

Catheter patency

Central venous catheters are commonly placed in cancer outpatients for the long-term administration of chemotherapeutic agents and fluids. The role of heparins in maintaining catheter patency has been examined by clinical trials and meta-analyses. A small, open-label study randomized 29 patients following placement of a subclavian venous catheter to dalteparin, 2500 IU daily for 90 days, or no prophylaxis [193]. Upper extremity DVTs were much less frequent in the dalteparin group (6% vs 62%, P = 0.002). A double-blind trial randomized 321 cancer patients to placebo or enoxaparin, 40 mg daily for 6 weeks [194]. DVTs were detected with similar frequency in the enoxaparin and placebo groups (14.1% and 18%), and there were no major bleeding events. Another double-blind study randomized 113 patients with active malignancies to placebo or nadroparin, 2850 IU daily for 3 weeks after central venous catheter placement [195]. No between-group differences were observed in the incidence of DVTs, catheter-related infections, or bleeding. Although an international group of experts recommended against routine prophylaxis for central venous catheters [196], a subsequent meta-analysis that included 12 randomized trials encompassing 3018 patients suggested that anticoagulation (warfarin or LMWH) significantly reduced the risk of symptomatic VTE from 6.8% to 3.7% (P < 0.001) [197].

During hemodialysis, fibrin forms in the extracorporeal circuit and dialyzer if antithrombotic measures are not implemented. Maintaining plasma heparin levels ≥ 0.5 IU/mL inhibits fibrin formation and the platelet release reaction [198]. A bolus dose of UH, 2000−4000 IU, given at the start of the dialysis session, is followed by up to 2000 IU/hour; an alternative is a bolus dose of 50 IU/kg and a continuous infusion of 800−1500 IU/hour [199]. LMWHs have many favorable characteristics compared with UH, but because most are eliminated by the kidney, they might bio-accumulate if used in hemodialysis patients. This hypothesis was refuted by a systematic review and meta-analysis of 17 studies that compared LMWH with UH in hemodialysis; the relative risk for total bleeding was 0.76 (95% confidence intervals 0.26−2.22) [200]. The authors concluded that LMWH was at least as safe as UH for extracorporeal circuit anticoagulation in chronic hemodialysis. The ultimate choice of antithrombotic agent is usually based on individual patient risk factors for bleeding or thrombosis, and economic considerations.

Anticoagulants are essential for preventing thrombus formation in mechanical circulatory support devices [201]. These are used to pump blood during cardiopulmonary bypass, extracorporeal membrane oxygenation (ECMO), the total artificial heart, and venous access devices. UH is used for most of these applications and monitored with the activated clotting time. Alternative short-acting anticoagulants for patients unable to tolerate heparin are bivalirudin or argatroban.

Dose modification for obesity or renal impairment

The prophylactic doses of anticoagulants might require modification in patients at the extremes of body weight. A clinical trial randomized 91 obese patients with an average body mass index (BMI) of 37.8 kg/m^2 to enoxaparin, 40 mg or 60 mg daily [202]. More patients in the 40 mg dose group had less than normal anti-Xa levels (64% vs 36%, P < 0.001), and in patients with weights > 100 kg, more of those receiving 60 mg had normal anti-Xa activity (44% vs 9%, P = 0.009), without an increase in adverse events. A review of pooled data from 11 studies found suggestive evidence that weight-based or higher-than-usual fixed doses was more likely to achieve desired anti-Xa levels, but further research was required [203]. Another review concluded that enoxaparin, 30 mg daily, might benefit low-body-weight patients and 40 mg twice daily would be appropriate for those with BMI ≥ 40 kg/m^2; higher doses might be needed if the BMI was ≥ 50 kg/m^2 [204]. On the other hand, a fixed dose of dalteparin, 5000 IU daily, was associated with similar incidences of VTE and major bleeding in obese (1118) and non-obese patients participating in a clinical trial [205]. The decision of whether to escalate the prophylactic dose of LMWH in obese patients should be based on the perceived risk of thrombosis, patient propensity for bleeding, and experience with a particular antithrombotic agent.

LMWHs and fondaparinux are mainly excreted by the kidneys, raising concern about the accumulation of these agents in patients with impaired renal function. Investigators have conducted clinical trials to assess safety and need for dosing modification. As noted previously, a subgroup analysis of the PROTECT trial that compared dalteparin, 5000 IU daily, with UH in critically ill patients with severe renal dysfunction observed no increase in major bleeding in the dalteparin group [145]. Another clinical trial measured anti-Xa levels in 120 patients with a creatinine clearance <30 mL/minute and found that no patient receiving dalteparin, 5000 IU daily, had bioaccumulation of the drug [206]. A systematic review of 11 publications confirmed that dalteparin, as well as tinzaparin, did not accumulate in patients with renal insufficiency, but accumulation was observed with enoxaparin, bemiparin, and certoparin; dose reduction was recommended when administering prophylactic doses of these latter LMWHs [207]. In addition, another review suggests that laboratory monitoring is appropriate if the creatinine clearance is <30 mL/minute or renal impairment is moderate and prolonged LMWH administration is anticipated [208]. With regard to enoxaparin, the manufacturer recommends that the prophylaxis doses be reduced to 30 mg once daily in patients with severe renal failure. Fondaparinux has been evaluated in patients with renal impairment in a prospective cohort study of 206 acutely ill medical patients with a mean creatinine clearance of 33 mL/minute; they received a reduced prophylactic dose of fondaparinux, 1.5 mg/kg daily for up to 15 days [209]. VTE was infrequent; one patient had major bleeding and eight had clinically-relevant non-major bleeding, suggesting that the lower prophylactic dose of the drug is safe and effective in these patients. However, fondaparinux is contraindicated in patients with a creatinine clearance <30 mL/minute.

Pregnancy

Pregnancy is characterized by a 4- to 10-fold increased risk of thrombosis due to elevated levels of procoagulants (factors VIII and Von Willebrand factor, factors VII, X, and fibrinogen) and decreased physiologic anticoagulants and fibrinolytic activity [210]. Studies have been conducted to determine whether pregnancy increases the risk for thrombotic events in women with inherited thrombophilia or previous VTE. In women with a variety of thrombophilias, a meta-analysis of 9 studies (2526 participants) reported odds ratios for VTE during pregnancy; the highest ratio was for factor V Leiden homozygotes (34.4) and prothrombin 20210 homozygotes (26.4) [211]. The American College of Chest Physicians (ACCP) guidelines recommend antenatal prophylaxis for women with these thrombophilias and a family history of VTE in a first degree relative [212]. The risk of having another thrombosis during pregnancy was recorded in a prospective study of 125 women with a single previous VTE [213]. Recurrences occurred in only 2.4% and were limited to those women with a previous unprovoked VTE and/or thrombophilia (5.9%). Consequently, the ACCP guidelines recommend antepartum anticoagulant prophylaxis only in those women with a prior unprovoked or estrogen-related VTE [211].

The increased risk of pregnancy-related thrombosis extends into the postpartum period. A study of claims data for the first recorded delivery in more than 1.5 million women compared VTE incidence data for the postpartum period with the same period one year later [214]. The odds ratio for a VTE appearance was 12.1 in the first 6 weeks postpartum but only 2.2 in the following 6 weeks. Because of this high risk, the ACCP guidelines recommend anticoagulant prophylaxis during the first 6 weeks postpartum in women homozygous for factor V Leiden or prothrombin 20210 mutation irrespective of whether they have a positive family history of thrombosis, and includes those with other thrombophilias and a positive family history of thrombosis [211]. They also suggest postpartum anticoagulant prophylaxis for women with a history of VTE. The American Society of Hematology (ASH) 2018 Guidelines are in harmony with these recommendations [215].

A 1994 commentary in the British Journal of Obstetrics and Gynaecology concluded that LMWH be considered an alternative to UH in pregnant women requiring antithrombotic prophylaxis because it was equally safe and effective, appeared to posed less risk of heparin-induced thrombocytopenia (HIT) and osteoporosis, and was more convenient to administer [216]. Clinical experience with enoxaparin, dalteparin, and tinzaparin is consistent with the safety of these LMWHs in pregnancy [217−219]. Transplacental passage of these LMWHs, as detected by elevated anti-Xa activity in cord blood, has not been demonstrated [220−222]. Fondaparinux has been administered to women unable to tolerate heparins because of hypersensitivity or other reactions, and cord blood anti-Xa was elevated in four of five such women, although the levels were well below the concentration required for effective anticoagulation [223]. Small case series have reported that fondaparinux appears to be safe for the fetus [224,225].

A retrospective cohort study of low-dose LMWH in 82 pregnancies at intermediate to high risk of thrombosis reported that the incidence of VTE was 5.5% (1.8% antepartum and 7% postpartum), and postpartum hemorrhage occurred in 21.6% (severe in 9.1%) [226]. These less than ideal outcomes might be due to the difficulty of establishing an optimal prophylactic dose. This might be because of changing dose requirements as pregnancy progresses. A study of 179 women revealed that 79% required adjustment of their LMWH dose to maintain an anti-Xa value in the prophylaxis range of 0.2−0.6 IU/mL [227]. However, it has not been established that anti-Xa levels within this range throughout pregnancy will improve outcomes. A clinical trial to determine the optimal dose LMWH has been initiated but not yet published [228]. The 2012 ACCP guideline recommended prophylactic or intermediate dose[1] LMWH, but the more recent ASH guideline suggests prophylactic rather than intermediate dose LMWH [211,214]. These recommendations apply to antepartum prophylaxis; postpartum, a 6-week course of either prophylactic or intermediate dose LMWH is suggested. The intensity of therapy should be based on an evaluation of the competing risks of thrombosis and bleeding, and individualized to meet the goals and expectations of each patient.

The management of anticoagulant therapy around the time of delivery has been examined. Excessive bleeding has been encountered if subcutaneous UH injections are given within 24 hours of delivery [229]. Prophylactic doses of LMWH pose less of a risk of bleeding, and if the injection is withheld when labor contractions begin, excessive bleeding is usually avoided at the time of delivery. The 2018 ASH guidelines advise against scheduled delivery with discontinuation of prophylactic anticoagulation compared with allowing spontaneous labor [214]. If a woman desires epidural anesthesia for her delivery, a meeting of the patient, the anesthesiologist, and the obstetrician should be convened to determine the least amount of time without LMWH protection acceptable for safe catheter insertion. The long half-life of fondaparinux poses a bleeding risk with spinal or epidural injections when women taking this anticoagulant during pregnancy enter labor. If they wish to have neuraxial anesthesia, they should be switched to a LMWH one week prior to the expected date of delivery [230].

The utility of heparins in women undergoing assisted reproduction and in those with recurrent pregnancy failures has been evaluated. The former is associated with a 3-fold risk of VTE, but there is no clear evidence that prophylactic anticoagulation improves outcomes [231]. For example, a recent retrospective cohort study of 230 women with two or more unsuccessful *in vitro* fertilization cycles given LMWH or no anticoagulant found no statistically significant difference between the two groups in the rates of clinical

---

1. For example, enoxaparin, 40 mg twice daily.

pregnancy and miscarriage [232]. On the other hand, the incidence of thrombosis in women hospitalized because of severe ovarian hyperstimulation syndrome is increased, which might justify the administration of LMWH [233]. The 2018 ASH guidelines include suggestions that prophylactic doses of LMWH be given to women with the ovarian hyperstimulation syndrome but not to those undergoing assisted reproduction therapy [214].

The role of anticoagulants in women with recurrent pregnancy losses has been examined. Enoxaparin, 40 mg or 80 mg daily, was given to 50 pregnant women with inherited thrombophilias and repeated miscarriages [234]. Live births were significantly more frequent than previously observed in these women (75% vs 20%, P < 0.0001). To confirm and extend this study, the investigators conducted a prospective, open-label trial that randomized 183 women at 5–10 weeks of gestation to either 40 mg or 80 mg enoxaparin [235]. Compared to patients' historical rates of live births (28%), outcomes were significantly improved (84% with the 40 mg dose and 78% with the 80 mg dose), and there were no major bleeding events. However, when subsequent trials included control groups, no benefit of LMWH was observed [236,237], and a double-blind, placebo-controlled trial of 258 pregnant women with unexplained miscarriages reported that enoxaparin, 40 mg daily until 35 weeks of gestation, did not increase the live-birth rate [238]. A meta-analysis of 8 trials enrolling 963 pregnant women concluded that LMWH did not produce a statistically significant reduction in placenta-mediated complications including pregnancy losses [239]. Therefore, recurrent pregnancy losses are not an indication for LMWH treatment during pregnancy.

Pediatrics

While VTE is generally uncommon in childhood, its incidence is not inconsequential in hospitalized children; the rate was 58 per 10,000 hospital admissions in 2007 [240]. Thrombosis occurs most often in the setting of childhood cancer, with central venous access devices being the commonest single risk factor [241]. A review of published guidelines noted that prophylactic anticoagulants are recommended only for children at high-risk, and usually combined with mechanical prophylaxis [242]. The choice of UH or LMWH has not been established by randomized, controlled trials, and should probably be based on the experience of the clinician.

References

[1] Jorpes JE. Heparin: a mucopolysaccharide and an active antithrombotic drug. Circulation 1959;19:87–91.
[2] Murray DWG, Best CH. The use of heparin in thrombosis. Ann Surg 1938;108:163–77.
[3] Simpson K. Shelter deaths from pulmonary embolism. Lancet 1940;ii:744.

[4]   Zilliacus H. On the specific treatment of thrombosis and pulmonary embolism with antic-oagulants, with particular reference to the post-thrombotic sequelae. Acta Med Scand 1946;(Suppl. 171):51−171.

[5]   Anonymous. Vijay Kakkar. BMJ 2017;356:6852.

[6]   Kakkar VV, Spindler J, Flute PT, Corrigan T, Fossard DP, Crelin RQ, et al. Efficacy of low doses of heparin in prevention of deep-vein thrombosis after major surgery: a double-blind, randomized trial. Lancet 1972;300:101−6.

[7]   Kakkar VV, Corrigan TP, Fossard DP, Sutherland I, Shelton MG, Thirlwall J. Prevention of fatal postoperative pulmonary embolism by low doses of heparin. Lancet 1975;306:45−64.

[8]   Sharnoff JG. Prevention of sudden cardiopulmonary arrest in the perioperative period with prophylactic heparin. Lancet 1969;ii:292−3.

[9]   Sharnoff JG, DeBlasio G. Prevention of fatal postoperative thromboembolism by heparin prophylaxis. Lancet 1970;ii:1006−7.

[10] Nicolaides AN, Dupont PA, Desai S, Lewis JD, Douglas JN, Dodsworth H, et al. Small doses of subcutaneous sodium heparin in preventing deep venous thrombosis after major surgery. Lancet 1972;ii:890−3.

[11] Gallus AS, Hirsh J, Tuttle RJ, Trebilcock R, O'Brien SE, Carroll JJ, et al. Small subcuta-neous doses of heparin in prevention of venous thrombosis. N Engl J Med 1973;288:545−51.

[12] Salzman EW. Low-molecular-weight heparin: is small beautiful? N Engl J Med 1986;315:957−9.

[13] Sharnoff JG, Rosen RL, Sadler AH, Cea RJ, Palazzo PJ. Prophylaxis of postoperative embolism. Lancet 1972;ii:488.

[14] Sagar S, Nairn D, Stamatakis JD, Maffei FH, Higgins AF, Thomas DP, et al. Efficacy of low-dose heparin in prevention of extensive deep-vein thrombosis in patients undergoing total-hip replacement. Lancet 1976;1(7970):1151−4.

[15] Morris GK, Henry AP, Preston BJ. Prevention of deep-vein thrombosis by low-dose hepa-rin in patients undergoing total hip replacement. Lancet 1974;2(7884):797−800.

[16] Hampson WGJ, Harris FC, Lucas HK, Roberts PH, McCall IW, Jackson PC, et al. Failure of low-dose heparin to prevent deep-vein thrombosis after hip-replacement arthroplasty. Lancet 1974;ii:795−7.

[17] Sharnoff JG. Prevention of thromboembolism. Bull N Y Acad Med 1973;49(8):655−60.

[18] Leyvraz PF, Richard J, Bachmann F, Van Melle G, Treyvaud J-M, Livio J-J, et al. Adjusted versus fixed-dose subcutaneous heparin in the prevention of deep-vein thrombo-sis after total hip replacement. N Engl J Med 1983;309:954−8.

[19] Imperiale TF, Speroff T. A meta-analysis of methods to prevent venous thromboembolism following total hip replacement. JAMA 1994;271:1780−5.

[20] Levine MN, Planes A, Hirsh J, Goodyear M, Vochelle N, Gent M. The relationship between anti-factor Xa level and clinical outcome in patients receiving enoxaparine low molecular weight heparin to prevent deep vein thrombosis after hip replacement. Thromb Haemost 1989;62:940−4.

[21] Turpie AGG, Levine MN, Hirsh J, Carter CJ, Jay RM, Powers PJ, et al. A randomized controlled trial of a low-molecular-weight heparin (enoxaparin) to prevent deep-vein thrombosis in patients undergoing elective hip surgery. N Engl J Med 1986;315:925−9.

[22] Spiro TE, Johnson GJ, Christie MJ, Lyons RM, MacFarlane DE, Blasier RB, et al. Efficacy and safety of enoxaparin to prevent deep venous thrombosis after hip replace-ment surgery. Ann Intern Med 1994;121:81−9.

[23] Colwell Jr CW, Spiro TE, Trowbridge AA, Morris BA, Kwaan HC, Blaha JD, et al. Use of enoxaparin, a low-molecular-weight heparin, and unfractionated heparin for the prevention of deep venous thrombosis after elective hip replacement. JBJS 1994;(76-A):3−14.

[24] Menzin J, Richner R, Huse D, Colditz GA, Oster G. Prevention of deep-vein thrombosis following total hip replacement surgery with enoxaparin versus unfractionated heparin: a pharmacoeconomic evaluation. Ann Pharmacother 1994;28:271−5.

[25] Levine MN, Hirsh J, Gent M, Turpie AG, Leclerc J, Powers PJ, et al. Prevention of deep vein thrombosis after elective hip surgery. Ann Intern Med 1991;114:545−51.

[26] Hull RD, Pineo GF, Stein PD, Mah AF, MacIsaac SM, Dahl OE, et al. Timing of initial administration of low-molecular-weight heparin prophylaxis against deep vein thrombosis in patients following elective hip arthroplasty. Arch Intern Med 2001;161:1952−60.

[27] Bergqvist D, Benoni G, Bjorgell O, Fredin H, Hedlundh U, Nicolas S, et al. Low-molecular-weight heparin (enoxaparin) as prophylaxis against venous thromboembolism after total hip replacement. N Engl J Med 1996;335:696−700.

[28] Planes A, Vochelle N, Darman J-Y, Fagola M, Beilaud M, Huet Y. Risk of deep-venous thrombosis after hospital discharge in patients having undergone total hip replacement: double-blind randomized comparison of enoxaparin versus placebo. Lancet 1996;348:224−8.

[29] Anderson DR, O'Brien BJ, Levine MN, Roberts R, Wells PS, Hirsh J. Efficacy and cost of low-molecular-weight heparin compared with standard heparin for the prevention of deep vein thrombosis after total hip arthroplasty. Ann Intern Med 1993;119:1105−12.

[30] O'Brien BJ, Anderson DR, Goeree R. Cost-effectiveness of enoxaparin versus warfarin prophylaxis against deep-vein thrombosis after total hip replacement. CMAJ 1994;150:1083−90.

[31] Menzin J, Colditz GA, Regan MM, Richner RE, Oster G. Cost-effectiveness of enoxaparin vs low-dose warfarin the prevention of deep-vein thrombosis after total hip replacement surgery. Arch Intern Med 1995;155:757−64.

[32] Torholm C, Broeng L, Jorgensen PS, Bierregaard P, Josephsen L, Jorgensen PK, et al. Thromboprophylaxis by low-molecular-weight heparin in elective hip surgery. A placebo controlled study. J Bone Joint Surg Br 1991;73:434−8.

[33] Eriksson BI, Kalebo P, Anthmyr BA, Wadenvik H, Tengborn L, Risberg B. Prevention of deep-vein thrombosis and pulmonary embolism after total hip replacement. JBJS 1991; (73-A):484−93.

[34] Francis CW, Pellegrini Jr VD, Totterman S, Boyd Jr AD, Marder VJ, Liebert KM, et al. Prevention of deep-vein thrombosis after total hip arthroplasty. JBJS 1997;79-A:1365−72.

[35] Monreal M, Lafoz E, Navarro A, Granero X, Caja V, Caceres E, et al. A prospective double-blind trial of a low molecular weight heparin once daily compared with conventional low-dose heparin three times daily to prevent pulmonary embolism and venous thrombosis in patients with hip fracture. J Trauma 1989;29:873−5.

[36] Dahl OE, Andreassen G, Aspelin T, Muller C, Mathiesen P, Nyhus S, et al. Prolonged thromboprophylaxis following hip replacement surgery-results of a double-blind, prospective, randomized, placebo-controlled study with dalteparin (Fragmin®). Thromb Haemost 1997;77:26−31.

[37] Lassen MR, Borris LC, Anderson BS, Jensen HP, Skejo Bro HP, Andersen G, et al. Efficacy and safety of prolonged thromboprophylaxis with a low molecular weight heparin (dalteparin) after total hip arthroplasty-the Danish Prolonged Prophylaxis (DaPP) study. Thromb Res 1995;89:281−7.

[38] Sarasin FP, Bounameaux H. Out of hospital antithrombotic prophylaxis after total hip replacement: low-molecular-weight heparin, warfarin, aspirin or nothing? A cost-effectiveness analysis. Thromb Haemost 2002;87:586−92.

[39] Ricotta S, Iorio A, Parise P, Nenci GG, Agnelli G. Post discharge clinically overt venous thromboembolism in orthopaedic surgery patients with negative venography-an overview analysis. Thromb Haemst 1996;76:887−92.

[40] Ascani A, Radicchia S, Parise P, Nenci GG, Agnelli G. Distribution and occlusiveness of thrombi in patients with surveillance detected deep vein thrombosis after hip surgery. Thromb Haemost 1996;75:239−41.

[41] Alatri A, Iorio A, Agnelli G. Choice of end-points in assessing the efficacy of post-discharge prophylaxis for venous thromboembolism. Thromb Haemost 1998;79:234−43.

[42] Arnesen H, Dahl OE, Aspelin T, Seljeflot I, Kierulf P, Lyberg T. Sustained prothrombotic profile after hip replacement surgery: the influence of prolonged prophylaxis with dalteparin. J Thromb Haemost 2003;1:971−5.

[43] Sobieraj DM, Lee S, Coleman CI, Tongbram V, Chen W, Colby J, et al. Prolonged versus standard-duration venous thromboprophylaxis in major orthopedic surgery. Ann Intern Med 2012;156:720−7.

[44] Lassen MR, Borris LC, Christiansen HM, Boll KL, Eiskjaer SP, Nielsen BW, et al. Prevention of thromboembolism in 190 hip arthroplasties. Acta Orthop Scand 1991;62:33−8.

[45] Hull RD, Raskob G, Pineo G, Rosenbloom D, Evans W, Mallory T, et al. A comparison of subcutaneous low-molecular weight heparin with warfarin sodium for prophylaxis against deep-vein thrombosis after hip or knee implantation. N Engl J Med 1993;329:1370−6.

[46] Planes A, Samama MM, Lensing AWA, Buller HR, Barre J, Vochelle N, et al. Prevention of deep vein thrombosis after hip replacement. Comparison between two low-molecular weight heparins, tinzaparin and enoxaparin. Thromb Haemost 1999;81:22−5.

[47] Kakkar VV, Howes J, Sharma V, Kadziola Z. A comparative, double-blind, randomized trial of a new second generation LMWH (bemiparin) and UFH in the prevention of post-operative venous thromboembolism. Thromb Haemost 2000;83:523−9.

[48] Lassen MR, Fisher W, Mouret P, Agnelli G, George D, Kakkar A, et al. Semuloparin for prevention of venous lthromboembolism after major orthopedic surgery: results from three randomized clinical trials, SAVE-HIP1, SAVE-HIP2 and SAVE-KNEE. J Thromb Haemost 2012;10:822−32.

[49] Turpie AGG, Gallus AS, Hoek JA. A synthetic pentasaccharide for the prevention of deep-vein thrombosis after total hip replacement. N Engl J Med 2001;344:619−25.

[50] Lassen MR, Bauer KA, Eriksson BI, Turpie AGG, for the European Pentasaccharide Hip Elective Surgery Study (EPHESUS) Steering Committee. Postoperative fondaparinux versus preoperative enoxaparin for prevention of venous thromboembolism in elective hip-replacement surgery: a randomized double-blind comparison. Lancet 2002;359:1715−20.

[51] Turpie AGG, Bauer KA, Eriksson BI, Lassen MR, for the PENTATHLON 2000 Study Steering Committee. Postoperative fondaparinux versus postoperative enoxaparin for prevention of venous thromboembolism after elective hip-replacement surgery: a randomized double-blind trial. Lancet 2002;359:1721−6.

[52] Eriksson BI, Bauer KA, Lassen MR, Turpie AGG, for the Steering Committee of the Pentasaccaride in Hip-Fracture Surgery Study. Fondaparinux compared with enoxaparin for the prevention of venous thromboembolism after hip-fracture surgery. N Engl J Med 2001;345:1298−304.

[53] Eriksson BI, Lassen MR, for the PENTasaccharide in Hip-FRActure Surgery Plus (PENTHIFRA Plus) investigators. Duration of prophylaxis against venous thromboembolism with fondaparinux after hip fracture surgery. Arch Intern Med 2003;163:1337−42.

[54] Sullivan DD, Kwong L, Nutescu E. Cost-effectiveness of fondaparinux compared with enoxaparin as prophylaxis against venous thromboembolism in patients undergoing hip fracture surgery. Value Health 2006;9:68−76.

[55] Eriksson BI, Borris LC, Friedman RJ, Haas S, Huisman MV, Kakkar AK, et al. Rivaroxaban versus enoxaparin for thromboprophylaxis after hip arthroplasty. N Engl J Med 2008;358:2765−75.

[56] Kakkar AK, Brenner B, Dahl OE, Eriksson BI, Mouret P, Muntz J, et al. Extended duration rivaroxaban versus short-term enoxaparin for the prevention of venous thromboembolism after total hip arthroplasty: a double-blind, randomised controlled trial. Lancet 2008;372:31−9.

[57] Lassen MR, Gallus A, Raskob GE, Pineo G, Chen D, Ramirez LM, et al. Apixaban versus enoxaparin for thromboprophylaxis after hip replacement. N Engl J Med 2010;363:2487−98.

[58] Eriksson BI, Dahl OE, Rosencher N, Kurth AA, van Dijk CN, Frostick SP, et al. Dabigatran etexilate versus enoxaparin for prevention of venous thromboembolism after total hip replacement: a randomised, double-blind, non-inferiority trial. Lancet 2007;370:949−56.

[59] Stulberg BN, Francis CW, Pellegrini VD, Miller ML, Shull S, DeSwart R, et al. Antithrombin III/low-dose heparin in the prevention of deep-vein thrombosis after total knee arthroplasty. Clin Orthop Relat Res 1989;248:152−7.

[60] Parmet JL, Berman AT, Horrow JC, Harding S, Rosenberg H. Thromboembolism coincident with tourniquet deflation during total knee arthroplasty. Lancet 1993;341:1057−8.

[61] Leclerc JR, Geerts WH, Desjardins L, Jobin F, Laroche F, Delorme F, et al. Prevention of deep vein thrombosis after major knee surgery-a randomized, double-blind trial comparing a low molecular weight heparin fragment (enoxaparin) to placebo. Thromb Haemost 1992;67:417−23.

[62] Colwell CE, Spiro TE, Trowbrdge AA, Stephens JWG, Gardiner Jr GA, Ritter KW. Efficacy and safety of enoxaparin versus unfractionated heparin for prevention of deep vein thrombosis after elective knee arthroplasty. Enoxaparin clinical trials group. Clin Orthop Relat Res 1995;321:19−27.

[63] Leclerc JR, Geerts WH, Desjardins L, Laflamme GH, l'Esperance B, Demers C, et al. Prevention of venous thromboembolism after knee arthroplasty: a randomized, double-blind trial comparing enoxaparin with warfarin. Ann Intern Med 1996;124:619−26.

[64] Leclerc JR, Gent M, Hirsh J, Geerts WH, Ginsberg JS, for the Canadian Collaborative Group. The incidence of symptomatic venous thromboembolism during and after prophylaxis with enoxaparin. Arch Intern Med 1998;158:873−8.

[65] Hull RD. Thromboprophylaxis in knee arthroscopy patients: revisiting values and preferences. Ann Intern Med 2008;149:137−9.

[66] Levine MN, Gent M, Hirsh J, Weitz J, Turpie AG, Powers P, et al. Ardeparin (low-molecular-weight heparin) vs graduated compression stockings for the prevention of venous thromboembolism. A randomized trial in patients undergoing knee surgery. Arch Intern Med 1996;156:851−6.

[67] Heit JA, Berkowitz SD, Bona R, Cabanas V, Corson JD, Elliott CG, et al. Efficacy and safety of low molecular weight heparin (ardeparin sodium) compared to warfarin for the

prevention of venous thrombosis after total knee replacement surgery: a double-blind, dose −ranging study. Thromb Haemost 1997;77:32−8.

[68] Heit JA, Elliott CG, Trowbridge AA, Morrey BF, Gent M, Hirsh J, et al. Ardeparin sodium for extended out-of-hospital prophylaxis against venous thromboembolism after total hip or knee replacement. A randomized, double-blind, placebo-controlled trial. Ann Intern Med 2000;132:853−61.

[69] Hamulyak K, Lensing AWA, van der Meer J, Smid WM, van Ooy A, Hoek JA, et al. Subcutaneous low-molecular weight heparin or oral anticoagulants for the prevention of deep-vein thrombosis in elective hip and knee replacement? Thromb Haemost 1994;74:1428−31.

[70] Navarro-quilis A, Castellet E, Rocha E, Paz-Jimenez J, Planes A, for the bemiparin study group in knee arthroplasty. Efficacy and safety of bemiparin compared with enoxaparin in the prevention of venous thromboembolism after total knee arthroplasty: a randomized, double-blind clinical trial. J Thromb Haemost 2003;1:425−32.

[71] Howard AW, Aaron SD. Low molecular weight heparin decreases proximal and distal deep venous thrombosis following total knee arthroplasty. Thromb Haemost 1998;79:902−6.

[72] Bauer KA, Eriksson BI, Lassen MR, Turpie AGG. Fondaparinux compared with enoxaparin for the prevention of venous thromboembolism after elective major knee surgery. N Engl J Med 2001;345:1305−10.

[73] Turpie AG, Bauer KA, Eriksson BI, Lassen MR. Fondaparinux vs enoxaparin for the prevention of venous thromboembolism in major orthopedic surgery: a meta-analysis of 4 randomized double-blind studies. Arch Intern Med 2002;162:1833−40.

[74] Bounameaux H, Perneger T. Enoxaparin or fondaparinux for thrombosis prevention after orthopaedic surgery. Lancet 2002;360:1702.

[75] Colwell Jr CW, Kwong LM, Turpie AG, Davidson BL. Flexibility in administration of fondaparinux for prevention of symptomatic venous thromboembolism in orthopaedic surgery. J Arthroplasty 2006;21:36−45.

[76] Eriksson BI, Bauer KA, Lassen MR, Turpie AGG. Influence of the duration of fondaparinux (Arixtra®) prophylaxis in preventing venous thromboembolism following major orthopedic surgery. J Thromb Haemost 2003;1:383−4.

[77] Singelyn FJ, Verheyen CC, Piovella F, Van Aken HK, Rosencher N, EXPERT Study Investigators. The safety and efficacy of extended thromboprophylaxis with fondaparinux after major orthopedic surgery of the lower limb with or without a neuraxial or deep peripheral nerve catheter: the EXPERT Study. Anesth Analg 2007;105:1540−7.

[78] Gordois A, Posnett J, Borris L, Bossuyt P, Jonsson B, Levy E, et al. The cost-effectiveness of fondaparinux compared with enoxaparin as prophylaxis against thromboembolism following major orthopedic surgery. J Thromb Haemost 2003;1:2167−74.

[79] Lowe GDO, Sandercock PAG, Rosendaal FR. Prevention of venous thromboembolism after major orthopaedic surgery: is fondaparinux an advance? Lancet 2003;362:504−5.

[80] Lassen MR, Ageno W, Borris LC, Lieberman JR, Rosencher N, Bandel TJ, et al. Rivaroxaban versus enoxaparin for thromboprophylaxis after total knee arthroplasty. N Engl J Med 2008;358:2776−86.

[81] Lassen MR, Raskob GE, Gallus A, Pineo G, Chen D, Portman RJ. Apixaban or enoxaparin for thromboprophylaxis after knee replacement. N Engl J Med 2009;361:594−604.

[82] Lassen MR, Raskob GE, Gallus A, Pineo G, Chen D, Hornick P, et al. Apixaban versus enoxaparin for thromboprophylaxis after knee replacement (ADVANCE-2): a randomised double-blind trial. Lancet 2010;375:807−15.

[83] Eriksson BI, Dahl OE, Rosencher N, Kurth AA, van Dijk CN, Frostick SP, et al. Oral dabigatran etexilate vs. subcutaneous enoxaparin for the prevention of venous thromboembolism after total knee replacement: the RE-MODEL randomized trial. J Thromb Haemost 2007;5:2178−85.

[84] Neumann I, Rada G, Claro JC, Carrasco-Labra A, Thorlund K, Akl EA, et al. Oral direct factor Xa inhibitors versus low-molecular-weight heparin to prevent venous thromboembolism in patients undergoing total hip or knee replacement. Ann Intern Med 2012;156:710−19.

[85] Adam SS, McDuffie JR, Lachiewicz PF, Ortel TL, Williams Jr. JW. Comparative effectiveness of new oral anticoagulants and standard thromboprophylaxis in patients having total hip or knee replacement: a systematic review. Ann Intern Med 2013;159:275−84.

[86] Kapoor A, Ellis A, Shaffer N, Gurwitz J, Chandramohan A, Saulino J, et al. Comparative effectiveness of venous thromboembolism prophylaxis options for the patient undergoing total hip and knee replacement: a network meta-analysis. J Thromb Haemost 2017;15:284−94.

[87] Mahmoudi M, Sobieraj DM. The cost-effectiveness of oral direct factor Xa inhibitors compared with low-molecular-weight heparin for the prevention of venous thromboembolism prophylaxis in total hip or knee replacement surgery. Pharmacotherapy 2013;33:1333−40.

[88] Camporese G, Bernardi E, Prandoni P, Noventa F, Verlato F, Simioni P, et al. Low-molecular-weight heparin versus compression stockings for thromboprophylaxis after knee arthroscopy. Ann Intern Med 2008;149:73−82.

[89] Camporese G, Noventa F, Bernardi E. Is low-molecular-weight heparin suitable in all patients undergoing knee arthroplasty? Ann Intern Med 2008;149:687−8.

[90] van Adrichem RA, Nemeth B, Algra A, le Cessie S, Rosendaal FR, Schipper IB, et al. Thromboprophylaxis after Knee Arthroscopy and Lower-Leg Casting. N Engl J Med 2017;376:515−25.

[91] Falck-Ytter Y, Francis CW, Johanson NA, Curley C, Dahl OE, Schulman S, et al. Prevention of VTE in orthopedic surgery patients: antithrombotic therapy and prevention of thrombosis, 9th ed: American College of Chest Physicians Evidence-Based Clinical Practice Guidelines. Chest 2012;141(Suppl. 2):e278S−325.

[92] Moll S. After the fall-prophylaxis for all? N Engl J Med 2017;376:576−7.

[93] Berger RE, Pai M, Rajasekhar A. Thromboprophylaxis after knee arthroscopy. N Engl J Med 2017;376:580−3.

[94] Geerts WH, Code KI, Jay RM, Chen E, Szalai JP. A prospective study of venous thromboembolism after major trauma. N Engl J Med 1994;331:1601−6.

[95] Geerts WH, Jay RM, Code KI, Chen E, Szalai JP, Saibil EA, et al. A comparison of low-dose heparin with low-molecular-weight heparin as prophylaxis against venous thromboembolism after major trauma. N Engl J Med 1996;335:701−7.

[96] Kock H-J, Schmit-Neuerburg KP, Hanke J, Rudofsky G, Hirche H. Thromboprophylaxis with low-molecular-weight heparin in outpatients with plaster-cast immobilization of the leg. Lancet 1995;346:459−61.

[97] Lassen MR, Borris LC, Nakov RL. Use of the low-molecular-weight heparin reviparin to prevent deep-vein thrombosis after leg injury requiring immobilization. N Engl J Med 2002;347:726−30.

[98] Ettema HB, Kollen BJ, Verheyen PM, Buller HR. Prevention of venous thromboembolism in patients with immobilization of the lower extremities: a meta-analysis of randomized controlled trials. J Thromb Haemost 2008;6:1093−8.

[99]   Zee AA, van Lieshout K, van der Heide M, Janssen L, Janzing HM. Low molecular weight heparin for prevention of venous thromboembolism in patients with lower-limb immobilization. Cochrane Database Syst Rev 2017;8:CD006681.

[100] Samama CM, Lecoules N, Kierzek G, Claessens YE, Riou B, Rosencher N, et al. Comparison of fondaparinux with low molecular weight heparin for venous thromboembolism prevention in patients requiring rigid or semi-rigid immobilization for isolated non-surgical below-knee injury. J Thromb Haemost 2013;11:1833−43.

[101] Clagett GP, Anderson Jr FA, Levine MN, Saltzman EW, Wheeler HB. Prevention of venous thromboembolism. Chest 1992;102(Suppl):391S−407S.

[102] Kakkar VV, Cohen AT, Edmonson RA, Phillips MJ, Cooper DJ, Das SK, et al. Low molecular weight versus standard heparin for prevention of venous thrombosis after major abdominal surgery. Lancet 1993;341:259−65.

[103] Bergqvist D, Burmark US, Flordal PA, Frisell J, Hallbook T, Hedberg M, et al. Low molecular weight heparin started before surgery as prophylaxis against deep vein thrombosis: 2500 versus 5000 XaI Units in 2070 patients. Br J Surg, 82. 1995. p. 496−501.

[104] Rasmussen MS, Jorgensen LN, Wille-Jorgensen P, Nielsen JD, Horn A, Mohn AC, et al. Prolonged prophylaxis with dalteparin to prevent late thromboembolic complications in patients undergoing major abdominal surgery: a multicenter randomized open-label study. J Thromb Haemost 2006;4:2384−90.

[105] Gazzaniga GM, Angelini G, Pastorino G, Santoro E, Lucchini M, Dal Pra MLA, et al. Enoxaparin in the prevention of deep venous thrombosis after major surgery: multi-centric study. Int Surg 1993;78:271−5.

[106] Nurmohamed MT, Verhaeghe R, Haas S, Iriate JA, Vogel G, van Rij AM, et al. A comparative trial of a low molecular weight heparin (enoxaparin) versus standard heparin for the prophylaxis ofpostoperative deep vein thrombosis in general surgery. Am J Surg 1995;169:567−71.

[107] McLeod RS, Geerts WH, Sniderman KW, Greenwood C, Gregoire RC, Taylor BM, et al. Subcutaneous heparin versus low-molecular-weight heparin as thromboprophylaxis in patients undergoing colorectal surgery: results of the Canadian colorectal DVT prophylaxis trial: a randomized, double-blind trial. Ann Surg 2001;233:438−44.

[108] Kakkar AK, Agnelli G, Fisher W, George D, Lassen MR, Mismetti P, et al. SAVE-ABDO investigators. Preoperative enoxaparin versus postoperative semuloparin thromboprophylaxis in major abdominal surgery: a randomized controlled trial. Ann Surg 2014;259:1073−9.

[109] Turpie AG, Bauer KA, Caprini JA, Comp PC, Gent M, Muntz JE, et al. Fondaparinux combined with intermittent pneumatic compression vs. intermittent pneumatic compression alone for prevention of venous thromboembolism after abdominal surgery: a randomized, double-blind comparison. J Thromb Haemost 2007;5:1854−61.

[110] Agnelli G, Bergqvist D, Cohen AT, Gallus AS, Gent M, PEGASUS investigators. Randomized clinical trial of postoperative fondaparinux versus perioperative dalteparin for prevention of venous thromboembolism in high-risk abdominal surgery. Br J Surg 2005;92:1212−20.

[111] ENOXACAN Study Group. Efficacy and safety of enoxaparin versus unfractionated heparin for prevention of deep vein thrombosis in elective cancer surgery: a double-blind randomized multicenter trial with venographic assessment. Br J Surg 1997;84:1099−103.

[112]   Bergqvist D, Agnelli G, Cohen AT, Eldor A, Nilsson PE, Moigne-Amrani AL, et al. Duration of prophylaxis against venous thromboembolism with enoxaparin after surgery for cancer. N Engl J Med 2002;346:975−80.

[113]   Kakkar VV, Balibrea JL, Martinez-Gonzalez J, Prandoni P, CANBESURE Study Group. Extended prophylaxis with bemiparin for the prevention of venous thromboembolism after abdominal or pelvic surgery for cancer: the CANBESURE randomized study. J Thromb Haemost 2010;8:1223−9.

[114]   Simonneau G, Laporte S, Mismetti P, Derlon A, Samii K, Samama CM, et al. A randomized study comparing the efficacy and safety of nadroparin 2850 IU (0.3 mL) vs. enoxaparin 4000 IU (40 mg) in the prevention of venous thromboembolism after colorectal surgery for cancer. J Thromb Haemost 2006;4:1693−700.

[115]   Koch A, Bouges S, Ziegler S, Dinkel H, Daures JP, Victor N. Low molecular weight heparin and unfractionated heparin in thrombosis prophylaxis after major surgical intervention: update of previous meta-analyses. Br J Surg 1997;84(6):750−9.

[116]   Akl EA, Terrenato I, Barba M, Sperati F, Sempos EV, Muti P, et al. Low-molecular-weight heparin vs unfractionated heparin for perioperative thromboprophylaxis in patients with cancer: a systematic review and meta-analysis. Arch Intern Med 2008;168:1261−9.

[117]   Hata K, Kimura T, Tsuzuki S, Ishii G, Kido M, Yamamoto T, et al. Safety of fondaparinux for prevention of postoperative venous thromboembolism in urological malignancy: a prospective randomized clinical trial. Int'L J Urol 2016;23:923−38.

[118]   Song J, Xuan L, Wu W, Shen Y, Tan L, Zhong M. Fondaparinux versus nadroparin for thromboprophylaxis following minimally invasive esophagectomy: a randomized controlled trial. Thromb Res 2018;166:22−7.

[119]   Matar CF, Kahale LA, Hakoum MB, Tsolakian IG, Etxeandia-Ikobaltzeta I, Yosuico VE, et al. Anticoagulation for perioperative thromboprophylaxis in people with cancer. Cochrane Database Syst Rev 2018;7:CD009447.

[120]   Bergqvist D. Low molecular weight heparin for the prevention of venous thromboembolism after abdominal surgery. Br J Surg 2004;91:965−74.

[121]   Kulik A, Rassen JA, Myers J, Schneeweiss S, Gagne J, Polinski JM, et al. Comparative effectiveness of preventative therapy for venous thromboembolism after coronary artery bypass graft surgery. Circ Cardiovasc Interv 2012;5:590−6.

[122]   Ramos R, Salem BI, De Pawlikowski MP, Coordes C, Eisenberg S, Leidenfrost R. The efficacy of pneumatic compression stockings in the prevention of pulmonary embolism after cardiac surgery. Chest 1996;109:82−5.

[123]   Gould MK, Garcia DA, Wren SM, Karanicolas PJ, Arcelus JI, Heit JA, et al. Prevention of VTE in nonorthopedic surgical patients: Antithrombotic Therapy and Prevention ofThrombosis, 9th ed: American College of Chest Physicians Evidence-Based Clinical Practice Guidelines. Chest 2012;141(Suppl. 2):e227S−77S.

[124]   Ho KM, Bham E, Pavey W. Incidence of venous thromboembolism and benefits and risk of thromboprophylaxis after cardiac surgery: a systematic review and meta-analysis. J Am Heart Assoc 2015;4:e002652.

[125]   Cerrato D, Ariano C, Fiacchino F. Deep vein thrombosis and low-dose heparin prophylaxis in neurosurgical patients. Neurosurg 1978;49:378−81.

[126]   Agnelli G, Pidovella F, Buoncristiani P, Severi P, Pini M, D'Angelo A, et al. Enoxaparin plus compression stockings compared with compression stockings alone in the prevention of venous thromboembolism after elective neurosurgery. N Engl J Med 1998;339:80−5.

[127]   Dickinson LD, Miller LD, Patel CP, Gupta SK. Enoxaparin increases the incidence of postoperative intracranial hemorrhage when initiated preoperatively for deep venous thrombosis prophylaxis in patients with brain tumors. Neurosurg 1998;43:1074−81.

[128]   Macdonald RL, Amidei C, Baron J, Weir B, Brown F, Erickson RK, et al. Randomized, pilot study of intermittent pneumatic compression devices plus dalteparin versus intermittent pneumatic compression devices plus heparin for prevention of venous thromboembolism in patients undergoing craniotomy. Surg Neurol 2003;59:363−72.

[129]   Nurmohamed MT, van Riel AM, Henkens CMA, Koopman MMW, Que GTH, d'Azemar P, et al. Low molecular weight heparin and compression stockings in the prevention of venous thromboembolism in neurosurgery. Thromb Haemost 1996;75:233−8.

[130]   Collen JF, Jackson JL, Shorr AF, Moores LK. Prevention of venous thromboembolism in neurosurgery: a metaanalysis. Chest 2008;134:237−49.

[131]   Perry JR, Julian JA, Laperriere NJ, Geerts W, Agnelli G, Rogers LR, et al. PRODIGE: a randomized placebo-controlled trial of dalteparin low-molecular-weight heparin thromboprophylaxis in patients with newly diagnosed malignant glioma. J Thromb Haemost 2010;8:1959−65.

[132]   Spinal Cord Injury Thromboprophylaxis Investigators. Prevention of venous thromboembolism in the acute treatment phase after spinal cord injury: a randomized, multicenter trial comparing low-dose heparin plus intermittent pneumatic compression with enoxaparin. J Trauma 2003;54:1116−24.

[133]   Spinal Cord Injury Thromboprophylaxis Investigators. Prevention of venous thromboembolism in the rehabilitation phase after spinal cord injury: prophylaxis with low-dose heparin or enoxaparin. J Trauma 2003;54:1111−15.

[134]   Chiou-Tan FY, Garza H, Chan KT, Parsons KC, Donovan WH, Robertson CS, et al. Comparison of dalteparin and enoxaparin for deep venous thrombosis prophylaxis in patients with spinal cord injury. Am J Phys Med Rehabil 2003;82:678−85.

[135]   Samama MM, Cohen AT, Darmon J-Y, Desjardins L, Eldor A, Janbon C, et al. A comparison of enoxaparin with placebo for the prevention of venous thromboembolism in acutely ill medical patients. N Engl J Med 1999;341:793−800.

[136]   Gardlund B for the Heparin Prophylaxis Study Group. Randomised, controlled trial of low-dose heparin for prevention of fatal pulmonary embolism in patients with infectious disease. Lancet 1996;347:1357−61.

[137]   Halkin H, Goldberg J, Modan M, Modan B. Reduction of mortality in general medical in-patients by low-dose heparin prophylaxis. Ann Intern Med 1982;96:561−5.

[138]   Warlow C, Beattie AG, Terry G, Ogston D, Kenmure ACF, Douglas AS. A double-blind trial of low doses of subcutaneous heparin in the prevention of deep-vein thrombosis after myocardial infaction. Lancet 1973;ii:93436.

[139]   Kakkar AJ, Cimminiello C, Goldhaber SZ, Parakh R, Wang C, Bergmann J-F, et al. Low-molecular-weight heparin and mortality in acutely ill medical patients. N Engl J Med 2011;365:2463−72.

[140]   Laporte S, Liotier J, Bertoletti L, Kleber FX, Pineo GF, Chapelle C, et al. Individual patient data meta-analysis of enoxaparin vs. unfractionated heparin for venous thromboembolism prevention in medical patients. J Thromb Haemost 2011;9:464−72.

[141]   Hull RD, Schellong SM, Tapson VF, Monreal M, Samama MM, Nicol P, et al. Extended-duration venous thromboembolism prophylaxis in acutely ill medical patients with recently reduced mobility. Ann Intern Med 2010;153:8−18.

[142]   Yusen RD, Hull RD, Schellong SM, Tapson VF, Monreal M, Samama MM, et al. Impact of age on the efficacy and safety of extended-duration thromboprophylaxis in

medical patients. Subgroup analysis from the EXCLAIM randomized trial. Thromb Haemost 2013;110:1152−63.

[143] Leizorovicz A, Cohen AT, Turpie AGG, Olsson C-G, Vaitkus PT, Goldhaber SZ, et al. Randomized, placebo-controlled trial of dalteparin for the prevention of venous thromboembolism in acutely ill medical patients. Circulation 2004;110:874−9.

[144] PROTECT Investigators for the Canadian Critical Care Trials Group and the Australian and New Zealand Intensive Care Society Clinical Trials Group. Dalteparin versus unfractionated heparin in critically ill patients. N Engl J Med 2011;364:1305−14.

[145] Pai M, Adhikari NKJ, Ostermann M, Heels-Ansdell D, Douketis JD, Skrobik Y, et al. Low-molecular-weight heparin venous thromboprophylaxis in critically ill patients with renal dysfunction: A subgroup analysis of the PROTECT trial. PLoS One 2018;13: e0198285.

[146] Cohen AT, Davidson BL, Gallus AS, Lassen MR, Prins MH, Tomkowski W, et al. Efficacy and safety of fondaparinux for the prevention of venous thromboembolism in older acute medical patients: randomized placebo controlled trial. BMJ 2006;332:325−9.

[147] Lloyd NS, Douketis JD, Moinuddin I, Lim W, Crowther MA. Anticoagulant prophylaxis to prevent asymptomatic deep vein thrombosis in hospitalized medical patients: a systematic review and meta-analysis. J Thromb Haemost 2008;6:405−14.

[148] Goldhaber SZ, Leizorovicz A, Kakkar AK, Haas SK, Merli G, Knabb RM, et al. Apixaban versus enoxaparin for thromboprophylaxis in medically ill patients. N Engl J Med 2011;365:2167−77.

[149] Cohen AT, Spiro TE, Buller HR, Haskell L, Hu D, Hull RD, et al. Rivaroxaban for thromboprophylaxis in acutely ill medical patients. N Engl J Med 2013;368:513−23.

[150] Spyropoulos AC, Ageno W, Albers GW, Elliott CG, Halperin JL, Hiatt WR, et al. Rivaroxaban for thromboprophylaxis after hospitalization for medical illness. N Engl J Med 2018;379:1118−27.

[151] Cohen AT, Harrington RA, Goldhaber SZ, Hull RD, Wiens BL, Gold A, et al. Extended thromboprophylaxis with betrixaban in acutely ill medical patients. N Engl J Med 2016;375:534−44.

[152] Gibson CM, Korjian S, Chi G, Daaboul Y, Jain P, Arbetter D, et al. Comparison of fatal or irreversible events with extended-duration betrixaban versus standard dose enoxaparin in acutely ill medical patients: an APEX trial substudy. J Am Heart Assoc 2017;6(7).

[153] Yee MK, Nafee T, Daaboul Y, Korjian S, AlKhalfan F, Kerneis M, et al. Increased benefit of betrixaban among patients with a history of venous thromboembolism: a post-hoc analysis of the APEX trial. J Thromb Thrombolysis 2018;45:1−8.

[154] Liew AY, Piran S, Eikelboom JW, Douketis JD. Extended-duration versus short-duration pharmacological thromboprophylaxis in acutely ill hospitalized medical patients: a systematic review and meta-analysis of randomized controlled trials. J Thromb Thrombolysis 2017;43:291−301.

[155] Schunemann HJ, Curshman M, Burnett AI, Kahn SR, Beyer-Westendorf J, Spencer FA, et al. American Sociate of Hematology 2018 guidelines for the management of venous thromboembolism: prophylaxis for hospitalized and non-hospitalized medical patients. Blood Adv 2018.

[156] Khorana AA, Kuderer NM, Culakova E, Lyman GH, Francis CW. Development and validation of a predictive model for chemotherapy-associated thrombosis. Blood 2008;111:4902−7.

[157] Kristinsson SY, Fears TR, Gridley G, Turesson I, Mellqvist UH, Bjorkholm M, et al. Deep vein thrombosis after monoclonal gammopathy of undetermined significance and multiple myeloma. Blood 2008;112:3582−6.

[158] Meyer G, Marjanovic Z, Valche J, Lorcerie B, Gruel Y, Solal-Celigny P, et al. Comparison of low-molecular-weight heparin and warfarin for the secondary prevention of venous thromboembolism in patients with cancer. Arch Intern Med 2002;162:1729−35.

[159] Pelzer U, Opitz B, Deutschinoff G, Stauch M, Reitzig PC, Hahnfeld S, et al. Efficacy of prophylactic low-molecular weight heparin for ambulatory patients with advanced pancreatic cancer: outcomes from the CONKO-004 Trial. J Clin Oncol 2015;33:2028−34.

[160] Khorana AA, Francis CW, Kuderer NM, Carrier M, Ortel TL, Wun T, et al. Dalteparin thromboprophylaxis in cancer patients at high risk for venous thromboembolism: a randomized trial. Thromb Res 2017;151:89−95.

[161] Agnelli G, Gussoni G, Bianchini C, Verso M, Mandala M, Cavanna L, et al. Nadroparin for the prevention of thromboembolic events in ambulatory patients with metastatic or locally advanced solid cancer receiving chemotherapy: a randomized, placebo-controlled, double-blind study. Lancet Oncol 2009;10:943−9.

[162] Van Doormaal FF, Di Nisio M, Otten HM, Richel DJ, Prins M, Buller HR. Randomized trial of the effect of the low molecular weight heparin nadroparin on survival in patients with cancer. J Clin Oncol 2011;29:2071−6.

[163] Sanford D, Naidu A, Alizadeh N, Lazo-Langer A. The effect of low molecular weight heparin on survival in cancer patients: an updated systematic review and meta-analysis of randomized trials. J Thromb Haemost 2014;12:1076−85.

[164] Agnelli G, George DJ, Kakkar AK, Fisher W, Lassen MR, Mismetti P, et al. Semuloparin for thromboprophylaxis in patients receiving chemotherapy for cancer. N Engl J Med 2012;366:601−9.

[165] Khorana AA, Soff GA, Kakkar AK, Vadhan-Raj S, Riess H, Wun T, et al. Rivaroxaban for thromboprophylaxis in high-risk ambulatory patients with cancer. N Engl J Med 2019;380:720−8.

[166] Carrier M, Abou-Nassar K, Mallick R, Tagalakis V, Shivakumar S, Schattner A, et al. Apixaban to prevent venous thromboembolism in patients with cancer. N Engl J Med 2019;380:711−19.

[167] Van Matre ET, Reynolds PM, MacLaren R, Mueller SW, Wright GC, Moss M, et al. Evaluation of unfractionated heparin versus low-molecular-weight heparin and fondaparinux for pharmacologic venous thromboembolic prophylaxis in critically ill patients with cancer. J Thromb Haemost 2018;16:2492−500.

[168] Di Nisio M, Carrier M, Lyman GH, Khorana AA, Subcommittee on Haemostasis and Malignancy. Prevention of venous thromboembolism in hospitalized medical cancer patients: guidance from the SSC of the ISTH. J Thromb Haemost 2014;12:1746−9.

[169] Connors JM. Prophylaxis against venous thromboembolism in ambulatory patients with cancer. N Engl J Med 2014;370:2515−19.

[170] Li W, Garcia D, Cornell F, Gailani D, Laubach J, Maglio ME, et al. Cardiovascular and thrombotic complications of novel multiple myeloma therapies: a review. JAMA Oncol 2017;3:980−8.

[171] McCarthy ST, Turner JJ, Robertson D, Hawkey CJ, Macey DJ. Low-dose heparin as a prophylaxis against deep-vein thrombosis after acute stroke. Lancet 1977;ii:800−1.

[172] Hillbom M, Erila T, Sotaniemi K, Tatlisumak K, Sarna S, Kaste M. Enoxaparin vs heparin for prevention of deep-vein thrombosis in acute ischaemic stroke: a randomized, double-blind study. Acta Neurol Scand 2002;106:84−92.

[173] Sherman DG, Albers GW, Bladin C, Fleschi C, Gabbai AA, Kase CS, et al. The efficacy and safety of enoxaparin versus unfractionated heparin for the prevention of venous thromboembolism after acute ischaemic stroke (PREVAIL Study): an open-label randomized comparison. Lancet 2007;369:1347−55.

[174] Turpie AGG, Hull RD, Schellong SM, Tapson VF, Monreal M, Samama MM, et al. Venous thromboembolism risk in ischemic stroke patients receiving extended-duration enoxaparin prophylaxis: results from the EXCLAIM study. Stroke 2013;44:249−51.

[175] Prins MH, Gelsema R, Sing AK, van Heerde LR, den Ottolander GJH. Prophylaxis of deep venous thrombosis with a low-molecular-weight heparin (Kabi 2165/Fragmin®) in stroke patients. Haemostasis 1989;19:245−50.

[176] Sandset PM, Dahl T, Stiris M, Rostad B, Scheel B, Abildgaard U. A double-blind and randomized placebo-controlled trial of low molecular weight heparin once daily to prevent deep-vein thrombosis in acute ischemic stroke. Semin Thromb Hemost 1990;16 (Supple):25−33.

[177] Diener HC, Ringelstein EB, von Kummer R, Landgraf H, Koppenhagen K, Harenberg J, et al. Prophylaxis of thrombotic and embolic events in acute ischemic stroke with the low-molecular-weight heparin certoparin: results of the PROTECT Trial. Stroke 2006;37:139−44.

[178] Shorr AF, Jackson WL, Sherner JH, Moores LK. Differences between low-molecular-weight and unfractionated heparin for venous thromboembolism prevention following ischemic stroke: a metaanalysis. Chest 2008;133:149−55.

[179] Lederle FA, Zylla D, MacDonald R, Wilt TJ. Venous thromboembolism prophylaxis in hospitalized medical patients and those with stroke: a background review for an American College of Physicians Clinical Practice Guideline. Ann Intern Med 2011;155:602−15.

[180] Qaseem A, Chou R, Humphrey LL, Starkey M, Shekelle P, for the Clinical Guidelines Committee of the American College of Physicians. Venous thromboembolism prophylaxis in hospitalized patients: a clinical practice guideline from the American College of Physicians. Ann Intern Med 2011;155:625−32.

[181] Wurm G, Tomancok B, Nussbaumer K, Adelwohrer C, Holl K. Reduction of ischemic sequelae following spontaneous subarachnoid hemorrhage: a double-blind, randomized comparison of enoxaparin versus placebo. Clin Neurol Neurosurg 2004;106:97−103.

[182] Gibson CM, Chi G, Halaby R, Korjian S, Daaboul Y, Jain P, et al. Extended-duration betrixaban reduces the risk of stroke versus standard-dose enoxaparin among hospitalized medically Ill patients: an APEX trial substudy (acute medically Ill venous thromboembolism prevention with extended duration betrixaban). Circulation 2017;135:648−55.

[183] Sharnoff JG. Prevention of coronary-artery thrombosis by heparin prophylaxis. Lancet 1973;ii:1321.

[184] Neri Serneri GG, Rovelli F, Gensini GF, Pirelli S, Carnovali M, Fortini A, et al. Effectiveness of low-dose heparin in prevention of myocardial reinfarction. Lancet 1987; i:939−42.

[185] Edmondson RA, Cohen AT, Das SK, Wagner MB, Kakkar VV. Low-molecular weight heparin versus aspirin and dipyridamole after femoropopliteal bypass grafting. Lancet 1994;344:914−18.

[186] Samama CM, Gigou F, Ill P, for the Enoxart Study Group. Low-molecular-weight heparin vs unfractionated heparin in femorodistal reconstructive surgery: a multicenter open randomized study. Ann Vasc Surg 1995;9(Suppl):S45−53.

[187] Lindholt JS, Houlind K, Gottschalksen B, Pedersen CN, Ravn H, Viddal B, et al. Five-year outcomes following a randomized trial of femorofemoral and femoropopliteal bypass grafting with heparin-bonded or standard polytetrafluoroethylene grafts. Br J Surg 2016;103:1300−5.

[188] Turpie AG, Robinson JG, Doyle DJ, Mulji AS, Mishkel GJ, Sealey BJ, et al. Comparison of high-dose with low-dose subcutaneous heparin to prevent left ventricular mural thrombosis in patients with acute transmural anterior myocardial infarction. N Engl J Med 1989;320:352−7.

[189] White DC, Grines CL, Grines LL, Marcovitz P, Messenger J, Schreiber T. Comparison of the usefulness of enoxaparin versus warfarin for prevention of left ventricular mural thrombus after anterior wall acute myocardial infarction. Am J Cardiol 2015;115:1200−3.

[190] Kontny F, Dale J, Abildgaard U, Pedersen TR. Randomized trial of low molecular weight heparin (dalteparin) in prevention of left ventricular thrombus formation and arterial embolism after acute anterior myocardial infarction: the Fragmin in acute myocardial Infarction (FRAMI) Study. J Am Coll Cardiol 1987;30:962−9.

[191] Bastiany A, Grenier ME, Matteau A, Mansour S, Daneault B, Potter BJ. Prevention of left ventricular thrombus formation and systemic embolism after anterior myocardial infarction: a systematic literature review. Can J Cardiol 2017;33:1229−36.

[192] Kajy M, Shokr M, Ramappa P. Use of direct oral anticoagulants in the treatment of left ventricular thrombus: systematic review of current literature. Am J Ther 2019. Available from: https://doi.org/10.1097/MJT.0000000000000937.

[193] Monreal M, Alastrue A, Rull M, Mira X, Muxart J, Rosell R, et al. Upper extremity deep venous thrombosis in cancer patients with venous access devices-prophylaxis with a low molecular weight heparin (Fragmin). Thromb Haemost 1996;75:251−3.

[194] Verso M, Agnelli G, Bertoglio S, Di Somma FC, Paoletti F, Ageno W, et al. Enoxaparin for the prevention of venous thromboembolism associated with central vein catheter: a double-blind, placebo-controlled, randomized study in cancer patients. J Clin Oncol 2005;23:4057−62.

[195] Niers TMH, Di Nisio M, Klerk CPW, Baarslag HJ, Buller HR, Biemond BJ. Prevention of catheter-related venous thrombosis with nadroparin in patients receiving chemotherapy for hematologic malignancies: a randomized, placebo-controlled study. J Thromb Haemost 2007;5:1878−82.

[196] Debourdeau P, Farge D, Beckers M, Baglin C, Bauersachs RM, Brenner B, et al. International clinical practice guidelines for the treatment and prophylaxis of thrombosis associated with central venous catheters in patients with cancer. J Thromb Haemost 2013;11:71−80.

[197] D'Ambrosio L, Aglietta M, Grignani G. Anticoagulation for central venous catheters in patients with cancer. N Engl J Med 2014;371:1362−3.

[198] Ireland H, Lane DA, Curtis JR. Objective assessment of heparin requirements for hemodialysis in humans. J Lab Clin Med 1984;103:643−52.

[199] Shen JI, Winkelmayer WC. Use and safety of unfractionated heparin for anticoagulation during maintenance hemodialysis. Am J Kidney Dis 2012;60:473−86.

[200] Lazrak HH, Rene E, Elftouh N, Leblanc M, Lafrance JP. Safety of low-molecular-weight heparin compared to unfractionated heparin in hemodialysis: a systematic review and meta-analysis. BMC Nephrol 2017;18:187.

[201] Kreuziger LB, Massicotte MP. Adult and pediatric mechanical circulation: a guide for the hematologist. Hematology Am Soc Hematol Educ Program 2018;2018:507−15.

[202] Miranda S, Le Cam-Duchez V, Benichou J, Donnadieu N, Barbay V, Le Besnerais M, et al. Adjusted value of thromboprophylaxis in hospitalized obese patients: a comparative study of two regimens of enoxaparin: the ITOHENOX study. Thromb Res 2017;155:1−5.

[203] He Z, Morrissey H, Ball P. Review of current evidence available for guiding optimal enoxaparin prophylactic dosing strategies in obese patients-actual weight−based vs fixed. Crit Rev Oncol Hematol 2017;113:191−4.

[204] Sebaaly J, Covert K. Enoxaparin dosing at extremes of weight: literature review and dosing recommendations. Ann Pharmacother 2018;52:898−909.

[205] Kucher N, Leizorovicz A, Vaitkus PT, Cohen AT, Turpie AG, Olsson CG, et al. Efficacy and safety of fixed low-dose dalteparin in preventing venous thromboembolism among obese or elderly hospitalized patients: a subgroup analysis of the PREVENT trial. Arch Intern Med 2005;165:341−5.

[206] Douketis J, Cook D, Meade M, Guyatt G, Geerts W, Skrobik Y, et al. Prophylaxis against deep vein thrombosis in critically ill patients with severe renal insufficiency with the low-molecular-weight heparin dalteparin: an assessment of safety and pharmacodynamics: the DIRECT study. Arch Intern Med 2008;168:1805−12.

[207] Atiq F, van den Bemt PM, Leebeek FW, van Gelder T, Versmissen J. A systematic review on the accumulation of prophylactic dosages of low-molecular-weight heparins (LMWHs) in patients with renal insufficiency. Eur J Clin Pharmacol 2015;71:921.

[208] Nutescu EA, Spinler SA, Wittkowsky A, Dager WE. Low-molecular-weight heparins in renal impairment and obesity: available evidence and clinical practice recommendations across medical and surgical settings. Ann Pharmacother 2009;43:1064−83.

[209] Ageno W, Riva N, Noris P, Di Nisio M, La Regina M, Arioli D, et al. Safety and efficacy of low-dose fondaparinux (1.5 mg) for the prevention of venous thromboembolism in acutely ill medical patients with renal impairment: the FONDAIR study. J Thromb Haemost 2012;10:2291−7.

[210] Brenner B. Clinical management of thrombophilia-related placental vascular complications. Blood 2004;103:4003−9.

[211] Robertson L, Wu O, Langhorne P, Twaddle S, Clark P, Lowe GD, et al. Thrombophilia in pregnancy: a systematic review. Br J Haematol 2006;132:171−96.

[212] Bates SM, Greer IA, Middeldorp S, Veenstra DL, Prabulos AM, Vandvik PO. VTE, thrombophilia, antithrombotic therapy, and pregnancy: antithrombotic therapy and prevention of thrombosis, 9th ed: American College of Chest Physicians Evidence-Based Clinical Practice Guidelines. Chest 2012;141(Suppl. 2):e691S−736S.

[213] Brill-Edwards P, Ginsberg JS, Gent M, Hirsh J, Burrows R, Kearon C, et al. Safety of withholding heparin in pregnant women with a history of venous thromboembolism. N Engl J Med 2000;343:1439−44.

[214] Kamel H, Navi BB, Sriram N, Hovsepian DA, Devereux RB, Elkind MSV. Risk of a thrombotic event after the 6-week postpartum period. N Engl J Med 2014;370:1307−15.

[215] Bates SM, Rajasekhar A, Middeldorp S, McLintock C, Rodger MA, James AH, et al. American Society of Hematology 2018 guidelines for management of venous

thromboembolism: venous thromboembolism in the context of pregnancy. Blood Adv 2018;2:3317−59.

[216]   Nelson-Piercy C. Low molecular weight heparin for obstetric thromboprophylaxis. Br J Obstet Gynaecol 1994;101:6−8.

[217]   Nelson-Piercy C, Letsky EA, de Swiet M. Low-molecular-weight heparin for obstetric thromboprophylaxis: experience of sixty-nine pregnancies in sixty-one women at high risk. Am J Obstet Gynecol 1997;176:1062−8.

[218]   Hunt BJ, Doughty H-A, Majumdar G, Copplestone A, Kerslake S, Buchanan N, et al. Thromboprophylaxis with low molecular weight heparin (Fragmin) in high risk pregnancies. Thromb Haemost 1997;77:39−43.

[219]   Jorgensen M, Nielsen JD. The low-molecular-weight-heparin, tinzaparin, is effective and safe in the treatment and prophylaxis of venous thromboembolic disease during pregnancy. Blood 2004;104:1774 (abst).

[220]   Forestier F, Sole Y, Aiach M, Alhenc Gelas M, Daffos F. Absence of transplacental passage of Fragmin (Kabi) during the second and the third trimesters of pregnancy. Thromb Haemost 1992;67:180−1.

[221]   Dimitrakakis C, Papageorgiou P, Papageorgiou I, Antzaklis A, Sakarelou N, Michalas S. Absence of transplacental passage of the low molecular weight heparin enoxaparin. Haemostasis 2000;30:243−8.

[222]   Omri A, Delaloye F, Andersen H, Bachmann F. Low molecular weight heparin Novo (LHN-1) does not cross the placenta during the second trimester of pregnancy. Thromb Haemost 1989;61:55−6.

[223]   Dempfle C-E H. Minor transplacental passage of fondaparinux in vivo. N Engl J Med 2004;350:1914−15.

[224]   Winger EE, Reed JL. A retrospective analysis of fondaparinux versus enoxaparin treatment in women with infertility or pregnancy loss. Am J Reprod Immunol 2009;62:253−60.

[225]   De Carolis S, di Pasquo E, Rossi E, Del Sordo G, Buonomo A, Schiavino D, et al. Fondaparinux in pregnancy: could it be a safe option? A review of the literature. Thromb Res 2015;135:1049−51.

[226]   Roeters Van Lennep JE, Meijer E, Klumper FJCM, Middeldorp JM, Bloemenkamp KWM, Middeldorp S. Prophylaxis with low-dose low-molecular-weight heparin during pregnancy and postpartum: is it effective? J Thromb Haemost 2011;9:473−80.

[227]   Boban A, Paulus S, Lambert C, Hermans C. The value and impact of anti-Xa activity monitoring for prophylactic dose adjustment of low-molecular-weight heparin during pregnancy: a retrospective study. Blood Coagul Fibrinolysis 2017;28:199−204.

[228]   Bleker SM, Buchmüller A, Chauleur C, Ní Áinle F, Donnelly J, Verhamme P, et al. Low-molecular-weight heparin to prevent recurrent venous thromboembolism in pregnancy: rationale and design of the Highlow study, a randomised trial of two doses. Thromb Res 2016;144:62−8.

[229]   Anderson DR, Ginsberg JS, Burrows R, Brill-Edwards P. Subcutaneous heparin therapy during pregnancy: a need for concern at the time of delivery. Thromb Haemost 1991;65:248−50.

[230]   Elsaigh E, Thachil J, Nash MJ, Tower C, Hay CR, Bullough S, et al. The use of fondaparinux in pregnancy. Br J Haematol 2015;168:762−4.

[231]   Bates SM. Anticoagulation and in vitro fertilization and ovarian stimulation. Hematology Am Soc Hematol Educ Program 2014;2014:379−86.

[232] Siristatidis C, Dafopoulos K, Salamalekis G, Galazios G, Christoforidis N, Moustakarias T, et al. Administration of low-molecular-weight heparin in patients with two or more unsuccessful IVF/ICSI cycles: a multicenter cohort study. Gynecol Endocrinol 2018;34 (9):747—51.

[233] Rova K, Passmark H, Lindqvist PG. Venous thromboembolism in relation to in vitro fertilization: an approach to determining the incidence and increase in risk in successful cycles. Fertil Steril 2012;97:95—100.

[234] Brenner B, Hoffman R, Blumenfeld Z, Weiner Z, Younis JS. Gestational outcome in thrombophilic women with recurrent pregnancy loss treated by enoxaparin. Thromb Haemost 2000;83:693—7.

[235] Brenner B, Hoffman R, Carp H, Dulitsky M, Younis J. LIVE-ENOX Investigators. Efficacy and safety of two doses of enoxaparin in women with thrombophilia and recurrent pregnancy loss: the LIVE-ENOX study. J Thromb Haemost 2005;3:227—9.

[236] Clark P, Walkcr ID, Langhorne P, Crichton L, Thomson A, Greaves M, et al. Scottish Pregnancy Intervention Study (SPIN) collaborators. SPIN (Scottish Pregnancy Intervention) study: a multicenter, randomized controlled trial of low-molecular-weight heparin and low-dose aspirin in women with recurrent miscarriage. Blood 2010;115:4162—7.

[237] Martinelli I, Ruggenenti P, Cetin I, Pardi G, Perna A, Vergani P, et al. Heparin in pregnant women with previous placenta-mediated pregnancy complications: a prospective, randomized, multicenter, controlled clinical trial. Blood 2012;119:3269—75.

[238] Pasquier E, de Saint Martin L, Bohec C, Chauleur C, Bretelle F, Marhic G, et al. Enoxaparin for prevention of unexplained recurrent miscarriage: a multicenter randomized double-blind placebo-controlled trial. Blood 2015;125:2200—5.

[239] Rodger MA, Gris JC, de Vries JIP, Martinelli I, Rey É, Schleussner E, et al. Low-molecular-weight heparin and recurrent placenta-mediated pregnancy complications: a meta-analysis of individual patient data from randomised controlled trials. Lancet 2016;388:2629—41.

[240] Raffini L, Huang YS, Witmer C, Feudtner C. Dramatic increase in venous thromboembolism in children's hospitals in the United States from 2001 to 2007. Pediatrics 2009;124:1001—8.

[241] Massicotte MP, Dix D, Monagle P, Adams M, Andrew M. Central venous catheter related thrombosis in children: analysis of the Canadian Registry of Venous Thromboembolic Complications. J Pediatr 1998;133:770—6.

[242] Faustino EV, Raffini LJ. Prevention of hospital-acquired venous thromboembolism in children: a review of published guidelines. Front Pediatr 2017;5:9.

Chapter 4

Treatment of thrombosis

Venous thrombosis

Deep vein thrombosis

A report of anticoagulant therapy for thrombotic disease from 1954 is of historical interest [1]. Unfractionated heparin (UH) was injected intramuscularly in an initial dose of 150 mg and a vitamin K antagonist was given along with the first dose of UH and continued daily. Subsequent doses of UH, 50 mg, were given every 4 hours for a total of 4 doses, and hyaluronidase was included with the UH to decrease pain, local bruising, and improve drug absorption. Monitoring was not considered necessary with the doses of UH employed, but if monitoring was desired, the therapeutic zone of the whole blood clotting time was considered to be 15−25 minutes. This anticoagulation regimen was given to 1135 patients with venous thromboembolism (VTE); as compared to historical controls, freedom from pain declined from 35 days to 3−4 days and confinement in bed from 40 days to 5−7 days. If venous thrombosis was recognized and anticoagulant therapy initiated, only 3 patients died of pulmonary embolism (PE), compared to 45 fatal emboli in those with unrecognized thrombi. The authors also noted that delay in instituting heparin therapy was more likely to result in post-thrombotic sequelae. Reviewing this report, it is curious that the UH was given intramuscularly rather than intravenously or subcutaneously, which would have avoided the risk of hematoma formation and need for hyaluronidase. In addition, the UH was discontinued prematurely, before the concentrations of the vitamin-K dependent clotting factors (II, IX, and X) would have decreased sufficiently to prevent thrombus formation. Nevertheless, this report presaged the modern treatment of VTE in several respects and suggested the possibility of treating selected patients at home rather than in hospital.

These and other early studies of heparin's effectiveness were mostly comprised of case-series, but in 1960, Barritt and Jordan [2] published the results of a controlled clinical trial of the use of anticoagulants in 35 patients with PE. Patients were randomized using a blinded allocation system to UH, 10,000 U intravenously every 6 hours for 6 doses, plus a coumarin drug, or no treatment. No monitoring was performed. The data were analyzed one year after the start of the trial, and it was observed that deaths from PE were

The Heparins. DOI: https://doi.org/10.1016/B978-0-12-818781-4.00004-2

increased in the 19 untreated patients compared with the 16 treated (26% vs 0%; P = 0.036). Recurrent PE was also more common in the untreated patients (5 vs 0). The next 38 patients recruited to the trial were all given anticoagulants and no fatalities were recorded. This was a landmark study, showing that giving substantial doses of UH are effective in preventing recurrent PE. It also confirmed that short courses of UH are associated with a low risk of serious hemorrhage and might not require monitoring.

Guidelines for the use of UH published in 1967 recommended that the anticoagulant be administered by intermittent intravenous infusion in doses ranging from 20,000 to 60,000 IU every 24 hours [3]. It was suggested that therapy should begin with relatively high doses; the risk of bleeding would be reduced if the dose was decreased as soon as the patient had stabilized. In fact, practitioners were cautioned that starting with small doses would likely result in inadequate coverage during the first critical one to two days of treatment. Heparin infusions should continue for at least 5 days, the length of time required for concomitantly initiated coumarin to exert a measurable antithrombotic effect. Lastly, it was noted that the risk of bleeding was increased in thrombocytopenic patients, those taking aspirin, and by intramuscular injections.

In 1969, the coagulation expert Daniel Deykin wrote that heparin had little effect on preformed thrombi except to limit platelet accretion on the thrombus surface and distal propagation of the clot, preventing further thrombus formation while clot lysis and vessel recanalization were in progress [4]. He observed that there was no uniformity of opinion concerning the method of regulating the dose, with the consequence that initial doses were often insufficient and subsequent doses excessive. A study published in 1972 attempted to remedy this situation by investigating the activated partial thromboplastin time (aPTT) for controlling UH therapy, and concluded that values between 1.5 and 2.5 times the aPTT of pooled normal plasma prevented recurrent VTE and decreased the risk of bleeding [5]. Although monitoring heparin activity with the aPTT was an improvement, patient outcomes continued to be suboptimal because of errors in preparing and infusing heparin solutions, as well as failure to re-calibrate the therapeutic aPTT reference range when new thromboplastin reagents were introduced [6,7]. The therapeutic range for the aPTT should be equivalent to a heparin level (by protamine titration) of 0.2−0.4 IU/mL or an anti-Xa level of 0.3−0.7 IU/mL [8]. VTE recurrence is more frequent in individuals whose aPTT level is less than therapeutic 24 hours after heparin initiation [9]. The danger of inadequate dosing is illustrated by the following case-study published in the Annals of Internal Medicine: [10]

Mary, the wife of a physician, was hospitalized and treated with UH for a venographically-confirmed DVT. On the fifth hospital day, she experienced dyspnea, chest pain, and hemoptysis; a lung scan showed bilateral filling defects consistent with multiple PE. Her husband discovered that the physicians caring for Mary had never ordered sufficient UH to prolong her aPTT into the therapeutic range. This failure was likely due to the use of a fixed-dose protocol for UH administration: an initial intravenous bolus of 5000 IU, followed by a continuous infusion of 1000 IU/hour (total dose, 29,000 IU) often resulted in subtherapeutic heparin levels. A clinical trial that increased the dose to 40,000 IU in the first 24 hours recorded a low incidence of recurrent VTE and only a marginal increase in bleeding complications (12% vs 9%) [11]. A subsequent trial compared fixed doses of heparin, 5000 IU bolus and 1000 IU/hour, with doses adjusted using a weight-based nomogram of 80 IU/kg bolus and 18 IU/kg/hour infusion [12]. The rate of recurrent VTE was significantly lower with the weight-adjusted regimen (5% vs 25%, P = 0.02), with no significant increase in bleeding, and this became the accepted UH protocol for the management of venous thrombotic disease.

Although most clinicians preferred intravenous UH therapy for VTE treatment, subcutaneous injections at 12 hour intervals were an alternative. After an initial dose of 15,000 IU, subsequent doses were adjusted to prolong the aPTT to 1.5−2 times the control. A study that randomized 103 patients with acute proximal DVT to intravenous or subcutaneous UH (the mean dose was 29,000 IU in each group) found no significant differences in the frequencies of pulmonary emboli or hemorrhagic complications [13]. A subsequent meta-analysis of eight clinical trials comparing the two methods of UH administration calculated that the overall relative risk for extension or recurrence of VTE was significantly lower with subcutaneous versus intravenous UH (0.62; 95% CI, 0.39−0.98), with no increased risk of hemorrhage [14]. Subsequently, a weight-based algorithm was published that resulted in a therapeutic aPTT in 99% of patients within 48 hours, enhancing the efficacy of subcutaneous UH for the initial treatment of VTE [15].

Anticoagulant therapy for VTE typically began with UH and concomitant administration of a coumarin for long-term management. UH administration was required for at least 5 days, the time necessary for vitamin K-dependent procoagulants to decline to levels protective against thrombosis. A double-blind, randomized trial examined whether 5 days of UH therapy was adequate to prevent VTE recurrence [16]. The outcomes for thrombosis and bleeding were similar irrespective of whether the 199 participants received heparin for 5 or 10 days, justifying a shorter course of inpatient anticoagulant therapy and earlier hospital discharge. A decrease in hospital stay was cost-effective; one analysis estimated that shortening the duration of heparin therapy could save as much as $600 million annually in the United States [17]. These cost savings could be realized if heparin was given subcutaneously rather than intravenously, and the injections continued as an outpatient when the patient was stable, avoiding the use of a vitamin K antagonist entirely. A randomized trial conducted in 1982 showed that treatment with subcutaneous UH, injected every 12 hours in doses adjusted to prolong the mid-interval aPTT to 1.5 times the control value, was as effective as warfarin and associated with less bleeding (1.8% vs 17%, P = 0.008) [18]. Such a regimen might be appropriate for pregnant patients in whom coumarin exposure is contraindicated.

The development of low molecular weight heparins (LMWH) began in the 1970s, and the products came into general clinical use in the 1990s. The generic and proprietary names of LMWH and synthetic heparins are shown in Box 4.1. They were attractive because of their subcutaneous route of administration, excellent bioavailability, relatively long half-life, and a predictable response, so that dose-monitoring was not required for most patients. Clinical trials were conducted to determine if LMWH were as efficacious as UH for the treatment of acute VTE. One such study was a double-blinded trial that randomized 432 patients with acute proximal vein thrombosis to **tinzaparin**, 175 IU/kg subcutaneously once daily, or UH, 5000 IU bolus followed by a continuous intravenous infusion of 40,320 IU/24 hours, adjusted daily to maintain the aPTT test within the therapeutic range [19]. Recurrent thromboses were infrequent in both groups, (LMWH 2.8% and UH 6.9%, P = 0.07) but major bleeding occurred less often with LMWH (0.5% vs 5%, P = 0.006), and there were fewer deaths in the LMWH group (4.7% vs 9.6%, P = 0.049). The authors observed that outpatient management of individuals with uncomplicated DVTs would be feasible with the simplified LMWH regimen.

Other LMWHs have been compared with UH for DVT treatment. A prospective open study randomized 204 patients to **dalteparin**, 200 IU/kg by daily subcutaneous injection, or continuous intravenous infusion UH, dose-adjusted to prolong the aPTT to the therapeutic range [20]. A venogram performed after 5 days showed no evidence of thrombus progression in either group, and no differences in VTE recurrence rates or major bleeding events. Another study in 153 patients provided confirmatory data [21]. Dalteparin was also compared with a coumarin for the management of acute VTE in 672 patients with cancer [22]. After initial treatment for 5–7 days with

---

BOX 4.1 Generic and proprietary names of low molecular weight and synthetic heparins

Generic name	Proprietary (Trade) name
Low Molecular Weight Heparins	
Dalteparin	Fragmin
Enoxaparin	Lovenox
Nadroparin	Fraxiparine
Tinzaparin	Innohep
Synthetic Heparin	
Fondaparinux	Arixtra

dalteparin, 200 IU/kg daily and a courmarin, patients were either continued on the coumarin alone for 6 months or were assigned to dalteparin, 200 IU for 1 month followed by 150 IU/kg for 5 months. Recurrent VTE occurred in fewer patients receiving dalteparin (8% vs 15.8%, P = 0.002), with no difference in the rate of major bleeding or mortality. Dalteparin is currently approved for the extended treatment of VTE in cancer patients in the doses used in this trial.

Enoxaparin has also been compared with UH for the management of DVT. An open-label study randomized 500 patients to enoxaparin, 1 mg/kg twice daily, or UH, 5000 IU intravenous bolus followed by a continuous infusion with the dose adjusted to maintain the aPTT within the therapeutic range [23]. No differences were observed in the rates of recurrent VTE (5% and 6%) or major bleeding, and patients assigned to the LMWH spent a mean of only 1.1 days in the hospital, compared to 6.5 days for the UH group. Another trial compared enoxaparin, 1 mg/kg twice daily or 1.5 mg/kg once daily, with UH in 900 patients with acute VTE [24]. Equivalent efficacy with both enoxaparin groups and UH was reported with low rates of recurrent symptomatic thrombosis ($\sim 4\%$) and major hemorrhage (2%). Based on this trial, enoxaparin was FDA-approved for inpatient VTE treatment with either 1.5 mg/kg once daily or 1.0 mg/kg every 12 hours in conjunction with warfarin; enoxaparin is continued for a minimum of 5 days and stopped when the INR is in the therapeutic range (INR, 2−3). For initial outpatient treatment of those with DVT but not PE, enoxaparin, 1 mg/kg every 12 hours, is given along with warfarin and continued for a minimum of 5 days and stopped when the INR is in the therapeutic range (INR, 2−3).

Another LMWH, **nadroparin**, was compared with UH for the treatment of VTE. In doses adjusted for body weight, this LMWH was found to be at least as effective as UH and outpatient use feasible and safe [25]. To determine whether most individuals with acute DVT would be eligible for home treatment, a Canadian study recruited 233 consecutive patients after excluding those with active bleeding, major PE (hemodynamic instability and requiring oxygen), pain requiring narcotics, or other indications for hospitalization [26]. The investigators found that 83% could be treated at home; the overall rates of VTE, bleeding, and death were comparable to those previously reported for hospitalized patients. A meta-analysis of trials published through 1995 concluded that LMWHs, administered subcutaneously in fixed doses adjusted only for body weight and without laboratory monitoring, were more effective and safer than adjusted-dose UH [27]. Four years later, another analysis confirmed that LMWHs were not only as safe and effective as UH, but also reduced mortality rates [28]. From an economic perspective, the investigators conducting the tinzaparin trial described previously calculated that the LMWH treatment was less expensive than UH, with a cost-savings of $40,149 per 100 treated patients [29]. A more comprehensive cost-effectiveness analysis using data from several randomized trials concluded that LMWHs were highly cost-effective for the inpatient management of VTE and had the potential for substantial cost savings when used for outpatient management [30].

Although intravenous UH has generally been replaced by LMWHs or direct oral anticoagulants for the treatment of acute VTE, fixed-dose, weight-adjusted subcutaneously injected UH is an alternative. A double-blind, non-inferiority trial randomized 708 patients to UH, first dose 333 U/kg and subsequent doses 250 U/kg, or LMWH, 100 U/kg; both given subcutaneously twice daily for ≥ 5 days until concomitantly administered warfarin became therapeutic [31]. No differences were observed in the frequencies of recurrent VTE, major bleeding, or death. Subcutaneous heparin might be a cost-saving treatment for thrombosis in resource-limited countries.

A synthetic heparin, **fondaparinux**, has also been evaluated for the treatment of symptomatic DVT. A double-blind trial randomized 2205 patients to subcutaneous injections of fondaparinux, 7.5 mg (± 2.5 mg for patients <50 kg or >100 kg) daily, or enoxaparin, 1 mg/kg twice daily, for 5 days; a vitamin K antagonist was given concomitantly and the primary efficacy outcomes were recurrent VTE, major bleeding, and death at 3 months [32]. Analysis of the results showed that fondaparinux was non-inferior to enoxaparin, and it was approved by the FDA for the treatment of acute symptomatic DVT.

Pulmonary embolism

Because the subcutaneous administration of LMWHs or fondaparinux enabled early ambulation of patients with acute DVTs, clinicians were concerned that thrombi in the leg veins might become detached and embolize to the lungs. This fear was put to rest by a study showing no increase in the prevalence of PEs in patients kept ambulatory during LMWH treatment of DVTs [33]. In addition, a clinical trial that randomized 129 patients to strict immobilization or daily ambulation reported new PEs in similar percentages of patients in the two groups (10% and 14%, P = 0.44), and no differences in the decrease of leg circumference or pain [34].

Although LMWHs had been found to be suitable for the treatment of DVT, there was uncertainty about their safety and effectiveness for acute symptomatic PE. These were addressed by a study that compared outcomes with tinzaparin and UH in 612 individuals with acute PE; excluded were patients with massive PE requiring thrombolytic therapy [35]. The composite endpoint of symptomatic recurrent VTE, major bleeding, or death was similar for patients on LMWH and UH: 3% and 2.9% by day 8 and 5.9% and 7.1% by day 90. Another study measured markers of coagulation activation and vascular obstruction scores in 37 patients with acute PE randomized to dalteparin or UH, and reported that the LMWH was at least as effective as UH in reducing clotting activity and perfusion abnormalities [36].

A meta-analysis of 12 randomized trials encompassing 1951 patients observed that there was no statistically significant difference in symptomatic VTE recurrence or major bleeding complications three months after treatment with LWMH or UH [37].

Fondaparinux was compared with UH for the initial treatment of PE in an open-label trial that randomized 2213 patients to daily fondaparinux, 7.5 ± 2.5 mg based on body weight, or continuous intravenous UH, dose-adjusted according to the aPTT [38]. A vitamin K antagonist was given concomitantly and the three-month primary outcome was a composite of symptomatic recurrent PE or DVT. The percentages of patients with recurrent VTE, major bleeding, or death were similar with the two agents, leading to FDA approval of fondaparinux for the initial treatment of acute symptomatic PE. The study investigators reported that 14.5% of those receiving fondaparinux were managed as outpatients.

The safety and feasibility of outpatient LMWH treatment for selected patients with PE was investigated by a study of 108 individuals not requiring hospitalization for reasons of hemodynamic instability, hypoxia, or severe pain; only 5.6% had symptomatic VTE recurrences and 1.9% major hemorrhages [39]. To confirm whether outpatient was non-inferior to inpatient management for PE, investigators conducted an open-label, randomized trial in 344 patients [40]. Non-inferiority for recurrent VTE within 3 months, death, and major bleeding was established, and the duration of hospitalization was reduced from a mean of 3.9 days to 0.5 days. Enthusiasm for outpatient treatment of PE is tempered by the recognition that patients selected for this approach meet criteria for being low-risk; the HESTIA criteria (Table 4.1) have been found to identify patients that can be safely treated at home [41].

TABLE 4.1 Hestia criteria for selection of patients with pulmonary embolism for home treatment. If any questions are answered in the affirmative, the patient is not eligible for outpatient management.

1. Hemodynamically unstable?
2. Thrombolysis or embolectomy necessary?
3. Active bleeding or high risk of bleeding?
4. Oxygen supply to maintain oxygen saturation >90% for >24 h?
5. Pulmonary embolism diagnosed during anticoagulant treatment?
6. Intravenous pain medication >24 h?
7. Medical or social reason for treatment in the hospital >24 h?
8. Creatinine clearance <30 mL/min?
9. Severe liver impairment?
10. Pregnant?
11. Documented history of heparin-induced thrombocytopenia?

Anticoagulant therapy for VTE is usually given for at least three months but might continue indefinitely depending on the risks of recurrence or bleeding [42]. Individuals whose thrombi were associated with a reversible risk factor such as surgery or trauma are at low risk of recurrence while those with unprovoked VTE might re-thrombose when anticoagulants are discontinued (the risk for fatal PE is 0.19−0.49 events per 100 patients completing a course of anticoagulants for a first VTE [43]). The relatively long duration of treatment for most patients favors the use of oral rather than injectable anticoagulants, but LMWHs have been given safely for up to 6 months in patients with contraindications to therapy with vitamin K antagonists [44]. Currently, most such patients would be treated with direct oral anticoagulants.

LMWH and fondaparinux are usually given in fixed doses and without monitoring. There are potential exceptions: pregnancy, pediatric use, obesity, and renal impairment. Pregnancy and pediatric use will be reviewed later in this chapter; obesity and impaired kidney function will be discussed in the following sections.

Obesity

The effect of body weight on the anticoagulant response to dalteparin, 200 IU/kg subcutaneously daily, was examined by measuring anti-Xa levels immediately before and 3−4 hours after injections on day 3 [45]. Peak and trough levels were similar in individuals weighing within 20%, 20%−40%, or >40% of ideal body weight (the largest patient weighed 190 kg). The safety of dosing dalteparin based on actual body weight was confirmed by a review of the VTE therapy of 193 obese patients weighing >90 kg; all received dalteparin, 200 IU/kg for 5−7 days followed by warfarin for three months, and only two had major bleeding, at 4 and 8 weeks from diagnosis [46].

The pharmacokinetics of enoxaparin were examined in obese and non-obese volunteers. After subcutaneous injection, steady-state exposure was achieved after the second dose in the non-obese and the third dose in the obese. Peak anti-Xa activity was similar in both groups, but the time to reach the peak was 1 hour longer in the obese individuals [47]. Another study observed that anti-Xa levels exceeding the therapeutic range (0.5−1.1 IU/mL) occurred in 50.5% of 99 morbidly obese patients receiving therapeutic doses of enoxaparin, although none had major bleeding [48]. Furthermore, an analysis of obese and non-obese individuals participating in three large clinical trials that included enoxaparin observed no difference in clinical outcomes that included the incidence of major bleeding [49,50].

Tinzaparin doses of 75 IU/kg and 175 IU/kg were evaluated in 37 individuals weighing from 101 kg to 165 kg [51]. Pharmacodynamics, as measured by anti-Xa and anti-IIa activities, were not influenced by body weight or body mass index. Also, 496 patients with weights >100 kg receiving therapeutic doses of weight-adjusted fondaparinux had similar rates of VTE recurrence and major bleeding as non-obese individuals [50]. In summary, these studies of three LMWHs and fondaparinux suggest that the dosing of obese individuals need not be capped at a maximal absolute dose, a position consistent with the current conditional recommendation of the guideline panel of the American Society of Hematology [52].

Renal impairment

Clinical trials have investigated whether LMWH therapy is safe in patients with kidney dysfunction. Prophylactic doses of dalteparin (5000 IU daily for 7 days) did not result in drug accumulation in patients with severe renal failure (mean creatinine clearance 19 mL/minute) [53]. A retrospective cohort study reported that the incidence of major bleeding in patients with chronic kidney disease was less in patients receiving therapeutic doses of dalteparin than in those treated with UH (1.1% vs 3.5%, P < 0.001) [54]. An analysis of data from a study that randomized 162 cancer patients with renal impairment to dalteparin 200 IU/kg daily for one month followed by 150 IU/kg during months 2–6, or warfarin for 6 months, observed similar bleeding event rates, but fewer VTE with the long-term dalteparin than with warfarin (2.7% vs 17%, P = 0.01) [22,55]. These studies suggest that dalteparin is relatively safe for patients with renal impairment.

Other clinical trials evaluated the effect of tinzaparin, 175 IU/kg daily for 5 days in >100 elderly patients with a mean creatinine clearance of 40 mL/minute [56,57]. A progressive increase in anti-Xa or anti-IIa levels was not observed, suggesting that systematic anti-Xa monitoring in these patients was not required. Furthermore, the rates of recurrent VTE and clinically relevant bleeding were similar in the tinzaparin and UH groups participating in the trial [58]. On the other hand, the ESSENCE and TIMI 11B trials reported that patients with severe renal impairment receiving enoxaparin experienced more major hemorrhages than those on UH (6.6% vs 1.1%, P < 0.001) [49]. A meta-analysis of 4 trials using standard enoxaparin dosing observed that peak anti-Xa levels were significantly higher in patients with creatinine clearances ≤ 30 mL/minute and major bleeding was increased (odds ratio 3.88) [59]. The manufacturer recommends that when using enoxaparin for VTE treatment in patients with severe renal impairment, the dose be decreased to 1 mg/kg daily. Treatment doses of nadroparin (180 IU/kg) administered to elderly individuals with modest renal impairment also led to anti-Xa accumulation [60].

The elimination of fondaparinux is principally by the kidney; 77% is excreted unchanged in the urine [61]. The clearance rate of the drug is 40% lower in patients with moderate renal failure, and these individuals had elevated bleeding rates when given prophylactic or therapeutic doses of the anticoagulant. Furthermore, the effects of the drug persist for several days because of its relatively long half-life (17−21 hours if renal function is normal). Therefore, fondaparinux is contraindicated in patients with severe renal failure and used with caution in those with moderate renal failure. In addition, the manufacturer recommends that in patients undergoing prolonged treatment, creatinine levels be assessed periodically and the drug discontinued if renal failure develops.

Direct oral anticoagulants are an alternative to LMWH in patients with kidney dysfunction; a clinical trial randomized patients with varying degrees of renal impairment to rivaroxaban, 15 mg twice daily for 3 weeks followed by 20 mg daily, or enoxaparin/vitamin K antagonist [62]. Major bleeding was more frequent in the enoxaparin/vitamin K antagonist recipients than with rivaroxaban (hazard ratios were 0.44 for mild and 0.23 for moderate renal impairment). These studies suggest that for the treatment of VTE in patients with mild to moderate renal impairment, a LMWH with minimal anti-Xa accumulation should be selected, and if such a LMWH is unavailable, a direct oral anticoagulant might be a safe alternative.

Venous thromboembolism in cancer

The increased risk of VTE in patients with malignancies can be attributed to multiple factors inherent in the tumor cells, their environment, and their treatment [63]. Anticoagulation with UH has been replaced by LMWH for a number of reasons including a decreased risk of recurrences, major bleeding, and death [64]. In addition, an *in vitro* study showed that the anticoagulant activity of UH, and to a lesser extent LMWH, was neutralized by necrotic tumor cell lines; the UH was bound by histones and ribosomal proteins exposed by the cells and was unable to potentiate antithrombin's inactivation of thrombin [65]. Clinical trials determined that continuation of LMWH therapy led to equivalent or better outcomes than switching to a vitamin K antagonist [19,66]. The LMWH is continued for as long as the malignant disorder is active [67], but substituting a vitamin K antagonist after 6 months of LMWH therapy is an alternative. A registry review showed that the risk of venous thrombosis was not increased in patients in whom this strategy was implemented [68]. Currently, dalteparin is FDA-approved for the extended (six months) treatment of VTE in patients with cancer; recommended doses are 200 IU/kg daily for the first month and 150 IU/kg daily for months 2 through 6.

Guidance for the management of incidental VTE in cancer patients has been provided by the International Society of Thrombosis and Haemostasis (ISTH) [69]. Therapeutic anticoagulation with LMWH for six months is recommended in all patients with incidental proximal DVT and PE with the exception of subsegmental PE without concomitant DVT. Clinical monitoring and serial ultrasound examinations are suggested for patients with subsegmental PEs and distal DVTs not receiving anticoagulants. However, a recent study showed that even with treatment, recurrence rates are high (6%) and similar whether the PE is proximal or subsegmental [70]. Periodic evaluation for bleeding risks and VTE recurrence should guide decisions about extending anticoagulation beyond six months.

Another issue addressed by the ISTH is the use of anticoagulants in patients with cancer-associated thrombosis and thrombocytopenia [71]. They suggest stratification based on the risk of recurrent thrombosis and the severity of the thrombocytopenia. The most difficult treatment decisions relate to patients with acute thrombosis (<30 days old) and platelet counts $<50 \times 10^9 \, L^{-1}$; a strategy for these individuals might be full-dose LMWH along with platelet transfusions to achieve higher platelet counts, or giving only half-dose or prophylactic doses of LMWH. Patients with higher platelet counts can tolerate full-dose LMWH, and those with persistent thrombocytopenia and thrombi present for more than 30 days have a reduced risk of recurrence and can be given smaller doses of anticoagulants.

VTE recurrence during anticoagulant therapy has been analyzed by a review of an international registry [72]. The authors noted that most treatment breakthroughs occurred while patients were receiving therapeutic doses of LMWH (70%); the dose was unchanged in 33%, increased in 31%, and switched to another drug in 24%. Additional recurrences were less likely with LMWH than a vitamin K antagonist (hazard ratio 0.28) and the bleeding rate was not increased. An alternative might be a direct oral anticoagulant; these new agents have been shown to be more effective than LMWH in reducing the incidence of VTE but increase the risk of major bleeding [73]. For example, hemorrhages occurred twice as often with edoxaban as with dalteparin (6.9% vs 4%, P = 0.04), and were mainly limited to individuals with gastrointestinal tumors, suggesting caution in the use of edoxaban in patients with this tumor-type [74]. Another trial reported that VTE recurrence rates were lower with rivaroxaban than with dalteparin (4% vs 11%) but clinically relevant non-major bleeding was more common (13% vs 4%) [75]. The ISTH recommends direct oral anticoagulants for cancer patients with acute VTE, a low risk of bleeding, and no drug-drug interactions with their current therapies [76].

The advantages and disadvantages of LMWH and direct oral anticoagulants in patients with cancer-associated VTE are listed in Table 4.2. As noted above, some of the direct oral anticoagulants are more effective than LMWH but are associated with more major bleeding [73]. Although some LMWHs accumulate in patients with renal failure, direct oral anticoagulants require only modest or no adjustment for creatinine clearances of 30−50 mL/minute. On the other hand, direct oral anticoagulants must be used with caution in patients with moderate liver impairment (Child-Pugh B). Bleeding might occur if direct oral anticoagulants are given concomitantly with chemotherapeutic agents that inhibit P-glycoprotein and CYP3A4 and conversely, their

TABLE 4.2 Features of low molecular weight heparins (LMWH) and direct oral anticoagulants (DOAC) relevant for the treatment of cancer-associated venous thromboembolism.

Feature	LMWH	DOAC
Efficacy	++	+++
Bleeding risk	++	+++
Accumulate in renal failure	++	+
Hepatic impairment	+	++
Drug-drug interactions[a]	0	++
HIT[b]	+	0
Route of administration	Injection	Oral

[a]P-glycoprotein and CYP3A4.
[b]HIT = heparin-induced thrombocytopenia.

anticoagulant action is weakened by medications that enhance these drug metabolizers [77]. Heparin-induced thrombocytopenia is a devastating but relatively infrequent complication of LMWH therapy (see Chapter 5). When selecting an anticoagulant, clinicians should elicit patient preferences with the goal of shared decision-making.

The safety and efficacy of long-term LMWH treatment (beyond six months) in patients with active malignancies and VTE has been examined. A prospective study examined outcomes in 334 patients receiving dalteparin for up to 12 months [78]. Major bleeding occurred in 3.6% in the first month and declined to 0.7% per patient-month by the end of the study; similarly, the incidence of recurrent VTE was 5.7% in the first month and declined to 4.1% during months 7−12. Another study evaluated tinzaparin in 247 patients and observed that the rate of clinically relevant bleeding was 0.6% per patient-month during months 7−12, but was significantly higher in men than women (hazard ratio 2.97, P = 0.02) [79]. The incidence of VTE recurrence was 1.1% during months 7−12. The conclusion from these two observational studies is that LMWH appears safe for long-term use in patients with cancer-associated VTE, but continued monitoring for new bleeding risk factors, especially in men, is mandatory.

Upper extremity deep vein thrombosis

Upper extremity DVT occasionally complicates the management of patients with central lines and underlying malignancy. A prospective cohort study of 46 outpatients showed that giving dalteparin, 200 IU/kg for 5 days, and warfarin for 3 months, was associated with only one recurrence and one major bleed [80]. Using the dalteparin/warfarin protocol, The Catheter Study reported no episodes of recurrent VTE and infrequent major bleeding (4%) in 74 patients; furthermore, no central lines were removed because of infusion failure or thrombus extension [81]. Another study compared dalteparin/warfarin with dalteparin alone in 67 patients and reported no VTE recurrences with either regimen on three month follow-up [82]. These studies informed the ISTH guidance for the management of catheter-associated DVT of the upper extremity in cancer patients [83]. Three to six months of LMWH therapy without removal of functioning catheters is recommended, and anticoagulation is suggested for incidental as well as symptomatic catheter-associated DVT.

Direct oral anticoagulants have been evaluated for central venous catheter-associated upper extremity DVT in cancer patients. A prospective cohort study administered rivaroxaban, 15 mg twice daily for three weeks, followed by 20 mg daily for 9 weeks, to 70 patients and reported VTE recurrences at 12 weeks in 1.43% and bleeding episodes in 12.9% [84]. One patient had a fatal PE while on treatment and the high bleeding rate was a concern. A recent review of the Swedish national anticoagulation registry reported that 55 patients with upper extremity DVT were given a single injection of LMWH followed by a direct oral anticoagulant (rivaroxaban in 84%) for 3−6 months [85]. The recurrence rate was 2% while on treatment and 4% after treatment was discontinued, and clinically relevant non-major bleeding occurred in only 2%. Because upper extremity thrombosis is relatively infrequent, formal clinical trials comparing LMWH with direct oral anticoagulants might not be undertaken and the optimal choice of anticoagulants remains unclear.

Distal deep vein thrombosis

The optimal management of distal DVT (calf-vein thrombosis) is controversial. Initially, it was assumed that all patients needed anticoagulant therapy, usually intravenous UH although an early trial showed that subcutaneous UH was no less effective [86]. Next, the requirement that UH be followed by three months of warfarin therapy in patients with symptomatic calf DVT was questioned, and it was observed that the VTE recurrence rate was significantly lower in patients receiving warfarin (0 vs 29%, P < 0.01) [87]. Subsequently, a prospective study that randomized 107 patients to compression therapy with or without nadroparin LMWH, 180 IU/kg daily, reported no difference in progression of thrombosis, vessel recanalization, PE, major bleeding, or death [88]. Another trial examining whether anticoagulant therapy (nadroparin 171 IU/kg daily) could be omitted in outpatients without active cancer or previous VTEs observed that the LMWH was not superior to placebo in preventing extension of DVT or VTE (3% vs 5%), but did increase the risk of bleeding (4% vs 0, P = 0.02) [89]. This was a randomized, double-blind study and all 252 patients wore compression stockings during the 6-week trial. A meta-analysis of 20 studies encompassing 2936 patients noted a reduction in the rate of recurrent VTE in patients receiving anticoagulants (odds ratio 0.50; 95% CI 0.31−0.79) without an increase in major bleeding [90]. Furthermore, the rate of PE was also lower (odds ratio 0.48, 0.25−0.91), and in four studies that compared 6 weeks with >6 weeks of anticoagulation, the longer course of treatment was associated with a lower risk of recurrent VTE (odds ratio 0.39, 0.17−0.90). The discrepant trial results are probably because of differences in the patient populations examined. An expert panel recommends anticoagulant therapy for isolated DVT in inpatients and those with severe symptoms, thrombi >5 cm in length, a history of VTE, or cancer; for most others, serial imaging of the deep veins for two weeks and anticoagulants only if the thrombus progresses in the calf veins or extends into the proximal veins [91]. Direct oral anticoagulants might have a role in the management of distal DVT, but clinical trial evidence of benefit and safety is not yet available.

Superficial venous thrombosis

Thrombi that form near the saphenofemoral junction are managed by saphenous vein ligation, thrombectomy, or anticoagulation. The treatment of superficial vein thrombosis distal to the saphenous venous junction has been the subject of a few clinical trials. A double-blind study randomized 164 patients with acute great saphenous vein thrombosis to a fixed dose of nadroparin, 2850 IU daily or a body-weight adjusted dose ($\sim$190 IU/kg) for 1 month [92]. No difference in thrombus progression or VTE complications was observed, and there was no major bleeding. To determine whether anticoagulation improved patient outcomes, a double-blind trial randomized 72 patients to dalteparin, initial dose 200 IU/kg followed by daily doses of 10,000 IU for 7 days, or ibuprofen, 800 mg three times daily for 14 days [93]. Thrombosis extension was less frequent in the dalteparin group (0 vs 11%, P = 0.05), and there were no bleeding events. Another double-blind trial randomized 3002 patients to fondaparinux, 2.5 mg daily, or placebo for 45 days [94]. The primary efficacy outcome of a composite of symptomatic VTE, extension to the saphenofemoral junction, or recurrent superficial vein thrombosis, occurred in fewer fondaparinux-treated patients (0.9% vs 5.9%, P < 0.001). The rate of PE was 85% lower with fondaparinux, and there was no increase in major bleeding. A Cochrane review concluded that fondaparinux was effective in preventing thrombus extension and VTE in superficial venous thrombosis, but LMWHs could not be evaluated because of the lack of high-quality evidence [95]. Fondaparinux was compared with the direct oral anticoagulant, rivaroxaban, in an open-label, non-inferiority phase 3b trial [96]. Patients (465) were randomized to fondaparinux, 2.5 mg subcutaneously, or rivaroxaban, 10 mg orally, for 45 days. No significant differences in the composite outcome of symptomatic DVT, PE, progression or recurrence of the superficial vein thrombosis, or death was observed and there were no major hemorrhages. The investigators concluded that rivaroxaban was non-inferior to fondaparinux and presented a less burdensome and expensive alternative therapy.

Retinal vein thrombosis

Anticoagulant therapy might be salutary for retinal vein thrombosis. A systematic review and meta-analysis of three clinical trials found that LMWH was associated with improvement in visual acuity, a 78% risk reduction for developing any adverse ocular outcome, and no increase in vitreous hemorrhages [97]. However, the role of LMWH in the beneficial outcomes was questioned because these trials did not compare LMWH with other contemporary ocular therapies such as lasers, glucocorticoids, and inhibitors of vascular endothelial growth factor, and they were not placebo-controlled [98]. Subsequently, investigators conducted a prospective cohort study; 111 patients with central or branch retinal vein occlusion were given nadroparin, 95 IU/kg twice daily, or enoxaparin, 1 mg/kg twice daily, followed by warfarin for 3 months [99]. Partial or complete recanalization of the retinal vein occlusion occurred in 70% and major bleeding in only 3.6%. Outcomes were better in younger patients and subsequent vascular events were more frequent in older patients (22% vs 5%, $P = 0.005$). A study of 13 patients with retinal vein occlusion receiving fondaparinux, 2.5 mg daily, observed that 9 of 11 continuing therapy had resolution of the venous occlusion and visual acuity improved in 7 [100]. In summary, the few reports in the literature suggest that anticoagulants are safe for patients with retinal vein occlusion, but further studies are needed to determine the circumstances under which they might be beneficial.

Cerebral sinovenous thrombosis

Thrombosis of the sagittal and other venous dural sinuses is suspected in a patient with headache, focal deficits, seizures, and papilledema [101], often in association with a history of oral contraceptive use. Initially, enthusiasm for treating this condition with anticoagulants was tempered by the occasional co-existence of intracranial bleeding. However, a clinical trial that randomized 20 patients to adjusted dose UH or placebo showed early improvement in the heparin group and complete recovery in 8 compared with 1 in the placebo group ($P < 0.01$) [102]. The investigators conducted a retrospective review of 102 patients and reported a decrease in mortality in those receiving UH (15% vs 69%). A placebo-controlled trial randomized 59 patients to nadroparin LMWH, 180 IU/kg daily for 3 weeks followed by a vitamin K antagonist for 3 months, or placebo [103]. Outcomes were similar in the two groups at 3 weeks, but at 12 weeks, fewer nadroparin-treated patients had a poor outcome (13% vs 21%); only one major bleed (non-intracranial) was reported. A non-randomized prospective cohort study that compared LMWH with UH observed that more patients receiving LMWH were functionally independent after 6 months (odds ratio, 2.4), and fewer had new intracerebral hemorrhages (odds ratio, 0.29) [104]. A trial that randomized 65 patients to dalteparin, 100 IU/kg twice daily, or adjusted dose UH, each for 14 days followed by 6 months of a vitamin K antagonist, reported a significant difference in mortality favoring the LMWH (0 vs 18.8%, P = 0.01), and no intracranial bleeding in either group [105]. Another small trial randomized 52 patients to enoxaparin, 1 mg/kg twice daily or adjusted dose UH, each for 7–10 days followed by a vitamin K antagonist [106]. Neurologic outcomes or mortality did not differ between the two groups, and there were no intracranial hemorrhages. The European Stroke Organization recommends LMWH for patients with acute cerebral venous thrombosis; [107] although intracranial bleeding is not considered a contraindication to anticoagulant therapy, a recent review suggests that half-doses of LMWH might be appropriate in the presence of a major hemorrhage [108].

Splanchnic venous thrombosis

The splanchnic vessels include the hepatic, portal, splenic, and mesenteric veins; patients with thrombosis of these veins often have an underlying cancer, myeloproliferative disorder, or inherited thrombophilia. Portal vein thrombosis is the most frequent and is usually associated with liver or pancreatic tumors, whereas myeloproliferative syndromes presage hepatic vein thrombosis (Budd-Chiari Syndrome) [109]. In 2012, the American College of Chest Physicians recommended anticoagulant therapy for patients with symptomatic, but not incidentally-detected, splanchnic vein thrombosis (SVT) [110]. However, a subsequent study that recorded outcomes in 177 patients with incidental SVT noted that the incidence of thrombotic events was similar whether patients had incidental or symptomatic thrombosis (8.0 events vs 7 events per 100 patient-years), suggesting that both groups receive anticoagulant therapy [111]. A review article commented that a LMWH or fondaparinux was particularly indicated for patients with extensive thrombosis of recent origin and thrombi that were progressing [112]. The authors suggested that parenteral therapy or a direct oral anticoagulant be continued for 3 or more months in patients with active cancers; in others, heparin injections could be replaced by a vitamin K antagonist or direct oral anticoagulant. A reduced dose of anticoagulant might be considered in patients with decreased platelet counts or other risk factors for bleeding. Anticoagulation could be discontinued after 3 months if the thrombosis was provoked by surgery or other transient risk factor; otherwise, therapy might continue indefinitely but the patient's risk for bleeding should be re-evaluated on a regular basis. A cohort study reported outcomes in 91 patients with portal vein thrombosis (hepatocellular carcinoma in 65%) treated with dalteparin or enoxaparin for six months [113]. Recanalization occurred in 61.5% and was more likely in patients with a favorable Child-Pugh class and more recent thrombosis. Bleeding was observed in 14.4%; risk factors were a history of variceal bleeding and low serum albumin. A prospective cohort study reported that 92 of 604 patients with splanchnic venous thrombosis were managed with anticoagulants for a median of 6.5 months [114]. Anticoagulation was associated with a 3.33-fold higher rate of vessel recanalization and a lower recurrent thrombosis rate, with no increase in major bleeding episodes. Although randomized trials have not been reported, clinical experience suggests that outcomes are improved by anticoagulant therapy and the risk of major hemorrhage mitigated by careful patient selection and the use of reduced doses as long as bleeding risk persists.

Renal vein thrombosis

Thrombosis of the renal veins is a consequence of selective permeability of the glomerular basement membrane, permitting loss of antithrombin and other proteins preventing thrombosis and accompanied by an increased synthesis of procoagulants (reviewed by Loscalzo [115]. Once the diagnosis is established, a LMWH is administered followed by a vitamin K antagonist [116]. If the risk factor for the renal vein thrombosis is reversible, 3−6 months of anticoagulant treatment is sufficient; otherwise, long-term anticoagulation is appropriate. Therapy with direct oral anticoagulants is a potential alternative that requires further evaluation.

Thrombosis of the renal veins is occasionally encountered in young people with anomalies of the inferior vena cava. These individuals usually develop signs and symptoms of lower extremity DVT and are often found to have thrombotic occlusion of the iliac veins [117]. Long-term anticoagulation is usually required because recurrent episodes of DVTs are not infrequent.

Venous thrombosis in pregnancy

Pregnancy is associated with an elevated risk for VTE that increases further during the puerperium [118]. Risk factors are cesarean section, inherited thrombophilia, older age, obesity, and postpartum infection [119,120]. The thrombi occur most frequently in the left leg and are more often proximal, associated with a greater risk of embolism and post-thrombotic syndrome. Following initial treatment with intravenous UH, it was previously recommended that UH be continued throughout the pregnancy rather than switch to a vitamin K antagonist. The latter endanger the fetus because they are teratogenic if administered during gestational weeks 6−12, and giving them during the third trimester increases the risk of fetal intracerebral hemorrhage during labor and delivery. LMWHs have replaced UH because of their longer plasma half-life, more predictable dose response, and simplicity of administration [121]. A systematic review and meta-analysis of 18 studies and 822 patients receiving LMWH confirmed the safety and efficacy of this approach; the weighted mean incidence of recurrent VTE during pregnancy was 1.97% and major bleeding occurred in 1.41% prior to delivery and 1.9% in the first 24 hours after delivery [122].

Two concerns with LMWH are maintenance of effective anticoagulation as pregnancy progresses, and management of dosing around the time of delivery. Because the volume of distribution of LMWH changes as weight increases during pregnancy, investigators suggested the LMWH dose be monitored with anti-Xa levels, with a target range of $0.5-1.2$ IU/mL $4-6$ hours after the morning dose [121]. However, clinical experience showed that anti-FXa monitoring was not required if the LMWH was adjusted for body weight in early pregnancy [123], and the American Society of Hematology 2018 Guidelines specifically recommend against routine monitoring of anti-FXa levels [124]. The suggested LMWH doses, based on body weight in early pregnancy, are dalteparin, 200 IU/kg once daily or 100 IU/kg twice daily; enoxaparin, 1.5 mg/kg once daily or 1 mg/kg twice daily; and tinzaparin, 175 IU/kg daily [118].

Pain during labor and delivery is usually managed with neuraxial anesthesia, but catheter insertion might trigger a spinal hematoma in an anticoagulated patient. To avoid such bleeding, LMWH injections are withheld for at least 24 hours before neuraxial blockade is performed [125]. Because the onset of labor is unpredictable, physicians and patients often pursue the alternative strategy of scheduling delivery and discontinuing LMWH 24 hours before induction of labor or performance of a cesarean section [118]. Postpartum, LMWH is not resumed for at least four hours after spinal anesthesia or removal of an epidural catheter, and patients are closely monitored for bleeding. Treatment is continued for at least 6 weeks, with a minimum total duration of three months. The concentration of LMWH or vitamin K antagonist in the breast milk of women receiving these anticoagulants is low and well-tolerated by newborns.

Pediatric venous thrombosis

Childhood VTE differs from adult VTE in incidence, etiology, and the frequency of complications, and its effective management requires recognition of these differences. Fortunately, recent publications from the American Society of Hematology address these issues and their recommendations with regard to the use of heparins will be briefly summarized here.

UH is still widely used in children for the treatment of VTE because concerns about the risk of anticoagulant-induced bleeding are mitigated by its short half-life and rapid onset/offset of action [126]. While the bolus and continuous intravenous infusion doses in older children are similar to those in adults (80 U/kg and 18 U/kg/hour), higher continuous infusion doses are required for infants (28 U/kg/hour) and young children (20 U/kg/hour). Monitoring of the aPTT or anti-Xa level is necessary but can be difficult because of limited venous access, occasionally requiring omission of the bolus dose and only gradual changes in the infusion rate. Of great concern is the potential for overdosing heparin if a vial with the wrong heparin concentration is infused; protamine should be available if rapid reversal of anticoagulation becomes necessary.

LMWHs have more predictable pharmacokinetics than UH and are given subcutaneously, which has popularized them for pediatric use; nevertheless, monitoring is required to ensure that therapeutic levels are achieved [127]. The therapeutic range is 0.5−1.0 anti-Xa U/mL in samples drawn 2−6 hours after injection. Treatment doses are similar to those for adults except that higher doses are needed for infants ≤ 2 months of age; they are 150 U/kg twice daily for dalteparin and 1.5 mg/kg twice daily for enoxaparin. Dalteparin has recently been FDA-approved for the treatment of VTE in pediatric patients; starting doses are based on patient age and weight, and the agent is continued for up to three months. Note that tinzaparin doses progressively decline from 275 U/kg to 175 U/kg daily in children ranging in age from ≤ 2 months to 16 years [126].

The American Society of Hematology guideline panel recommends that pediatric patients with symptomatic DVT or PE be given anticoagulants, but equivocate on the use of anticoagulants in those with asymptomatic VTE, suggesting that treatment decisions be individualized [128]. The panel suggests that anticoagulant therapy (either LMWH or warfarin) be continued for ≤ 3 months for patients with provoked DVT or PE and 6−12 months in those with unprovoked DVT or PE. Removal of non-functioning; thrombosed central vascular access devices be delayed for a few days after initiation of anticoagulant therapy to reduce the risk of embolization. They also suggest using anticoagulants in neonates with renal vein thrombosis and in pediatric patients with portal vein thrombosis associated with an occlusive thrombus. They strongly recommend the use of anticoagulants in pediatric patients with cerebral sinovenous thrombosis, and conditionally suggest the use of anticoagulants even in those with intracerebral hemorrhage.

Post-thrombotic syndrome

The post-thrombotic syndrome (PTS) refers to the development of pain, swelling, itching, and discoloration of the limb within the first 1−2 years after a DVT, accompanied by signs of venous hypertension and reflux [129]. Risk factors for PTS are extensive or recurrent DVT, obesity, and older age [130]. Another major risk factor is under-dosing of anticoagulants, principally vitamin K antagonists, during the initial weeks of acute DVT treatment [131]. Prolonged administration of therapeutic doses of LMWH mitigates this risk [132,133]. A pooled analysis of 5 studies noted a risk ratio favoring LMWH over vitamin K antagonist for complete recanalization of thrombosed veins (0.66, $P < 0.001$) and a lower incidence of PTS [134]. A modest decrease in PTS has also been achieved by taking advantage of the anti-inflammatory properties of statin drugs. A prospective trial that randomized 234 patients to bemiparin LMWH alone or with rosuvastatin reported lower levels of C-reactive protein and less PTS (38.3% vs 48.5%, $P = 0.019$) in the rosuvastatin group [135]. Other modalities for the management of the PTS are elastic compression hose, exercise, and lifestyle modifications, but whether long-term anticoagulation is beneficial is still under evaluation [130].

An alternative to warfarin is the direct oral anticoagulant, rivaroxaban; small clinical trials have suggested that in comparison with LMWH/warfarin, rivaroxaban is associated with a modest decrease in PTS incidence [136,137]. However, analysis of a national registry of 19,957 individuals with acute VTE treated with warfarin or rivaroxaban did not show a significant difference in PTS frequency (0.55 vs 0.53 incidents per 100 person-years) [138].

Proximal DVT is a strong risk factor for PTS [139], but whether anticoagulants should be combined with thrombus removal is uncertain. A clinical trial that randomized patients to anticoagulants with or without pharmacomechanical catheter-directed thrombolysis reported that at 2 years there was no difference in the incidence of PTS ($\sim 48\%$) but more major bleeding events with thrombolysis (1.7% vs 0.3%, $P = 0.049$) [140]. However, fewer patients in the thrombolysis group had developed moderate-to-severe PTS (18% vs 24%, $P = 0.04$), suggesting that patients with very severe DVT or iliofemoral vein thrombosis might be more likely to benefit from catheter-directed thrombolysis [130].

Arterial disease

The potential of heparins for the treatment of atherosclerosis and arterial thrombosis was recognized by investigators such as Engelberg, who championed the use of UH because of its lipid clearing activity, anti-proliferative properties, and ability to retard thrombus formation (see Chapter 2). Other scientists showed that LMWH as well as UH could reduce the deposition of platelets and fibrinogen on the denuded pig carotid artery and decrease restenosis in the hypercholesterolemic rabbit iliac artery [141,142]. Anticoagulants administered to patients with acute myocardial infarction (MI) prevented the development of mural thrombi and reduced stroke incidence [143]. LMWHs were introduced into clinical practice in the early 1990s and shortly thereafter were compared with UH in clinical trials of patients with angina and MI. The results of these and other studies will be discussed in the following paragraphs.

Acute coronary syndromes (unstable angina and myocardial infarction)

Clinical trials in the 1980s reported that aspirin was beneficial in patients with unstable angina, and that adding UH further improved outcomes [144−146]. On the other hand, the addition of UH to aspirin and thrombo-lytic therapy for acute MI provided only minor additional benefit and was not recommended [147]. Therefore, researchers sought to determine if the newly emerging LMWHs, which appeared to have a number of pharmaco-logical advantages over UH, might have a role in the treatment of unstable angina and acute MI.

In 1995, investigators conducted a randomized trial comparing nadroparin with UH in 211 patients with unstable angina receiving aspirin [148]. Recurrent angina was less frequent with LMWH than with UH (21% vs 44%, P = 0.003). Other studies showed that dalteparin was more effective than placebo in reducing the rate of new MI, revascularization, and death during the first 6 days of treatment (2.2% vs 5.7%, P < 0.001) [149]. While receiving twice-daily dalteparin during the acute phase of treatment, partici-pants had significantly decreased thrombin generation which increased when the dose of dalteparin was reduced to once-daily [150]. Investigators next compared dalteparin with UH in a randomized trial of 1482 patients with unstable angina or non-Q-wave MI [151]. During the first six days, patients were randomized to dalteparin, 120 IU every 12 hours, or UH; between days 6 and 45, they received dalteparin, 120 IU daily, or placebo. Differences in the rates of recurrent angina, MI, and death between dalteparin and UH were not significant during the acute phase of the trial, suggesting that the LMWH was an effective alternative to UH; prolonging dalteparin therapy beyond six days did not provide additional benefit over placebo. A second study ran-domized 2267 patients to placebo or twice daily, weight-adjusted dalteparin: 5000 IU for women < 80 kg and men < 70 kg, and 7500 IU for all others [152]. The composite endpoint of death, MI, or both was significantly lower in the dalteparin group (3.1% vs 5.9%, P < 0.002) at one month, and the dif-ference between groups was still significant at three months if the need for revascularization was an included outcome. However, major bleeding was more frequent in the dalteparin group (3.3% vs 1.5%). Dalteparin, 120 IU/kg every 12 hours (with aspirin) is FDA-approved for the prevention of ische-mic complications in patients with unstable angina and non-Q-wave MI.

Enoxaparin, 1 mg/kg twice daily for up to 8 days, was compared with UH in a randomized, double-blind trial that included 3171 patients with angina at rest or non-Q-wave MI [153]. The incidence of recurrent angina, MI, or death was significantly lower with enoxaparin than with heparin, both at day 14 and day 30 (16.6% vs 19.8%, $P = 0.019\%$, and 19.8% vs 23.3%, $P = 0.016$, respectively), and the need for revascularization was also less frequent (27% vs 32.2%, $P = 0.001$). Major bleeding was not increased and the benefits of enoxaparin were sustained for at least one year [154]. An examination of biomarkers found that the levels of Von Willebrand factor increased during the initial 48 hours and predicted the composite endpoint at both 14 and 30 days; the rise was significantly greater with UH than with enoxaparin [155]. An economic analysis found that the total medical costs were less for enoxaparin than heparin, with a total cumulative savings per patient at 30 days of $1172 ($P = 0.04$) [156].

Another trial randomized 3910 patients to UH or enoxaparin, 30 mg intravenous bolus followed by subcutaneous injections of 1 mg/kg every 12 hours; upon discharge, patients received placebo or enoxaparin, 40 mg if <65 kg and 60 mg if >65 kg for 35 days [157]. The primary endpoint of death, MI, urgent revascularization occurred in fewer patients in the enoxaparin group during the first 8 days (12.4% vs 14.5%, $P = 0.048$) and persisted through day 43 (17.3% vs 19.7%, $P = 0.048$). During the outpatient phase, patients receiving enoxaparin had significantly more major bleeding (2.9% vs 1.5%, $P = 0.021$). An open label trial examined the safety of administering enoxaparin in conjunction with eptifibatide, a glycoprotein IIb/IIIa inhibitor [158]. A total of 746 patients were randomized to receive enoxaparin, 1 mg/kg twice daily, or UH for 48 hours, and all were given aspirin and eptifibatide. Major bleeding was less with enoxaparin than with UH (1.8% vs 4.6%, $P = 0.03$) but minor bleeding was more frequent (30.3% vs 20.8%, $P = 0.003$). However, patients in the enoxaparin group were less likely to have ischemic electrocardiographic changes during monitoring (14.3% vs 25.4%, $P = 0.0002$), and fewer MIs or deaths (5% vs 9%, $P = 0.03$).

Trials also compared enoxaparin with UH when these heparins are used in combination with thrombolytic therapy for patients with acute MI. 6095 patients were randomized to full-dose tenecteplase with either enoxaparin or UH [159]. The endpoint, a composite of re-infarction, refractory ischemia, or death, occurred in significantly fewer patients in the enoxaparin group than in the UH group (11.4% vs 15.4%, $P = 0.0002$), with a non-significant increase in major bleeding (3% vs 2.2%). Enoxaparin was also compared with UH in a randomized trial of 20,506 patients with ST-elevation MI scheduled for thrombolytic therapy [160]. The composite of recurrent MI, death, or urgent revascularization occurred less often with enoxaparin than with UH (11.7% vs 14.5%, $P < 0.001$), but major bleeding was more frequent with enoxaparin (2.1% vs 1.4%, $P < 0.001$).

A meta-analysis conducted in 2000 concluded that LMWH and UH were associated with similar rates of MI, death, and bleeding when administered for the treatment of unstable angina or non-Q-wave MI [161], but another meta-analysis performed in 2007 that included patients with ST- and non-ST-segment elevation (a total of 49,088 patients in 12 trials) observed that the one month endpoint of death or MI was significantly reduced by enoxaparin compared with UH (9.8% vs 11.4%, P < 0.001) [162]. However, major bleeding was greater with enoxaparin than with UH (4.3% vs 3.4%, P = 0.019), and patients with ST-segment elevation benefitted more than those with non-ST-segment elevation. The FDA-approved dose of enoxaparin for patients with unstable angina and non-Q-wave MI is 1 mg/kg daily to prevent ischemic complications. Individuals with acute ST-segment elevation MI should receive an intravenous bolus dose of 30 mg plus a 1 mg/kg subcutaneous dose followed by 1 mg/kg every 12 hours, with dose adjustments for patients ≥ 75 years of age (no initial bolus and once daily dosing). All LMWH doses are administered in conjunction with aspirin.

The safety and efficacy of reviparin LMWH for the treatment of acute MI was evaluated in a trial that randomized 15570 patients to placebo or weight-adjusted reviparin every 12 hours for 7 days [163]. Study drugs were started along with thrombolytic therapy, and the composite endpoint was death, re-infarction, and stroke. At 7 days, fewer patients in the reviparin group had the composite endpoint (11% vs 9.6%, P = 0.005) but had more major bleeding (0.2% vs 0.1%, P = 0.07).

Fondaparinux was compared with enoxaparin in a dose-finding study of 1138 patients with acute coronary syndromes [164]. The lowest event rates were observed with fondaparinux doses of 2.5 mg (P < 0.05), and bleeding rates were low. A double-blind trial assigned 20,078 patients to fondaparinux, 2.5 mg daily or enoxaparin, 1 mg/kg twice daily, for six days and recorded death, MI, or refractory ischemia at 9 days [165]. No differences between groups in these outcomes were observed, but major bleeding was less frequent with fondaparinux (2.2% vs 4.1%, P = < 0.001). Furthermore, fondaparinux treatment was associated with fewer deaths at 30 days (P = 0.02). Subsequent analyses of this trial found that fondaparinux was associated with a lower mean anticoagulant intensity than enoxaparin, possibly accounting for less bleeding, and the decrease in major hemorrhages reduced the odds ratio for death three-fold [166,167]. Another trial in patients with acute ST-segment elevation MI reported that the composite endpoint of death or re-infarction at 9 days was significantly reduced by fondaparinux as compared with placebo (7.4% vs 8.9%, P = 0.003), and major bleeding was not increased (1.8% vs 2.1%) [168]. Fondaparinux compared favorably with UH when used as an adjunct to thrombolytic therapy for acute MI; the incidence of re-occlusion of the infarct-related vessel and need for re-vascularizations, as well as frequency of major bleeding, were not significantly different [169].

A direct oral anticoagulant, rivaroxaban, was studied in a double-blind, placebo-controlled trial of 15,526 patients with a recent acute coronary syndrome. A twice-daily dose of 2.5 mg was associated with a lower composite endpoint of cardiovascular death, MI, or stroke than placebo (9.1% vs

10.7%, P = 0.02), and although the rates of bleeding were increased (1.8% vs 0.6%, P < 0.001), fatal bleeding events occurred in only 0.1% [170]. A second trial randomized 27,395 participants with stable vascular disease to rivaroxaban, 2.5 mg twice daily plus aspirin, or aspirin alone [171]. The primary endpoint occurred in fewer patients in the rivaroxaban/aspirin group than in the aspirin alone group (4.1% vs 5.4%, P < 0.001), but there were more major bleeding events (3.1% vs 1.9%, P < 0.001). A systematic review and meta-analysis of 6 trials that included 29,667 patients receiving a combination of direct oral anticoagulant and antiplatelet therapy noted that the benefit of direct oral anticoagulants was confined to individuals with ST-segment elevation MI; no significant treatment effect was observed in patients with non-ST-elevation MI [172]. Direct oral anticoagulants were associated with a higher risk of major bleeding compared with antiplatelet therapy alone (odds ratio 3.17, P < 0.001). Rivaroxaban, 2.5 mg twice daily, combined with aspirin was approved by the FDA to reduce the risk of major cardiovascular events in people with chronic coronary or peripheral artery disease.

Percutaneous coronary intervention

LMWHs have also been used as adjuncts to percutaneous coronary intervention (PCI). A prospective, open-label study randomized 3528 patients to an intravenous bolus of enoxaparin, 0.5 or 0.75 mg/kg, or UH; the primary endpoint was bleeding and the secondary endpoint, achievement of target anticoagulation levels [173]. Less bleeding occurred with the 0.5 mg/kg dose than with UH (5.9% vs 8.5%, P = 0.01) and the target anticoagulation levels were reached in more patients (79% vs 20%, P < 0.001). A subsequent small study reported that death or recurrent MI in hospital and at one month was reduced in those receiving enoxaparin as compared to UH (odds ratios 0.28 and 0.35, respectively), and there were fewer bleeding events (1.2% vs 2.6%) [174]. A recent meta-analysis that included 8861 patients from 4 trials determined that the rates of death and MI, as well as major bleeding, were similar with enoxaparin and UH [175].

A pilot study of fondaparinux in 61 patients undergoing percutaneous coronary angioplasty showed inhibition of thrombin generation, acute vessel closure in only 3.3%, and no major bleeding, suggesting that further evaluation of the agent was warranted [176]. In a follow-up blinded-trial, 350 patients were randomized to UH or intravenous fondaparinux, 2.5 mg or 5.0 mg [177]. The incidence of total bleeding was similar in all groups, but less common with fondaparinux 2.5 mg than 5 mg. However, a larger trial failed to show a benefit of fondaparinux over UH because guiding catheter thrombosis was more frequent [168]. Laboratory experiments demonstrated that catheter segments triggered the plasma contact activation system, and that UH and enoxaparin, but not fondaparinux, attenuated this contact-induced clotting [178]. To prevent thrombi from forming in catheters, supplementation of fondaparinux with small amounts of UH was required. Therefore, fondaparinux is of limited value for the prevention or treatment of catheter-induced thrombosis.

Prosthetic heart valves/bridging

Long-acting oral anticoagulants are typically used to prevent thrombi from forming on prosthetic heart valves, but interruptions for invasive procedures or other reasons might necessitate temporary employment of a short-acting anticoagulant. Douketis [179] suggests using LMWH for bridging patients with prosthetic heart valves in the following circumstances: the appearance of a thrombotic event in a patient with a subtherapeutic INR, an anticipated prolonged interruption of warfarin therapy because of major surgery, and while awaiting warfarin to induce a therapeutic INR in a patient with a newly implanted prosthetic valve. Enoxaparin administered on the first postoperative day in 1063 patients undergoing mechanical heart valve replacement was complicated by thromboembolism in 1%, major bleeding events in 4.1%, and no deaths [180]. A systematic review and meta-analysis of 23 studies that included 9534 patients having mechanical valve replacement reported a low rate of thromboembolism (1.1%); however, the bleeding rate was lower with UH than with LMWH (2.2% vs 5.5%) [181].

Dosing guidelines from the American College of Chest Physicians advise stratification of patients according to the risks of thrombosis and bleeding; [182] for example, thrombotic risk is higher for patients with mechanical prosthetic valves in the mitral than aortic position [183]. The factors increasing the risk of major bleeding were identified by a double-blind, placebo-controlled trial of bridge therapy with dalteparin, 100 IU/kg twice daily in 1813 patients with atrial fibrillation [184]. They were perioperative aspirin use, a history of renal failure, a post-procedure international normalized ratio (INR) >3.0, and having a high-bleeding risk procedure. Another factor that might increase bleeding is insufficient delay between the last dose of LMWH and the start of surgery. Anti-Xa activity, measured 14 hours after dosing enoxaparin, 1 mg/kg twice daily, was reported as high as 0.5 IU/mL in 68% of patients [185] -a level that might enhance surgical bleeding.

Pregnancy in a patient with a prosthetic heart valve is another indication for an effective anticoagulant to replace coumarin derivatives such as warfarin in order to avoid fetal malformations and stillbirth. Warfarin exposure between 6 and 12 weeks of gestation is most likely to induce teratogenicity, but fetal embryopathy is unpredictable and can occur even with low doses of vitamin K antagonists [186]. The dose of the anticoagulant replacing warfarin must be sufficient to prevent thrombi from forming on the prosthetic valve; if a LMWH is selected, trough anti-Xa levels should be in the 0.7−1.2 IU/mL range [187]. An observational study of 58 women with mechanical valves reported death in 9%, serious maternal morbidity in 41%, and poor fetal outcome in 47% of the pregnancies [188]. A LMWH was used by 71% of these women throughout pregnancy and most required rapid dose escalation in the first trimester. An alternative is adjusted-dose UH given subcutaneously every 12 h to keep the mid-interval aPTT at least twice control or an anti-Xa heparin level of 0.35−0.70 IU/mL [189]. Clinical experience suggests that treatment should be individualized and patients have meticulous monitoring managed by skilled practitioners.

Ischemic stroke

Studies conducted in the 1980s failed to show a benefit of UH in the treatment of transient ischemic attacks or ischemic strokes [190−192]. But in 1995, a double-blind trial that randomized 2750 patients within 48 hours of symptom onset to nadroparin, 4100 IU twice or once daily for 10 days, or placebo, showed a dependency/mortality advantage for the LMWH (45% and 52% vs 65%, P = 0.005 for the trend) [193]. However, the benefits of LMWH were found to be comparable to those of aspirin when the two were evaluated in double-blind clinical trials. One study randomized 449 patients with acute ischemic stroke to dalteparin, 100 IU/kg twice daily, or aspirin, and reported no significant differences in the percentages of recurrent strokes, cerebral bleeding, functional outcomes, and death in the two groups at 14 days and three months [194]. A second study compared two doses of tinzaparin with aspirin in 1486 patients with acute ischemic stroke, and observed no differences between groups in the proportions of patients independent at 6 months [195]. A Cochrane Review noted that LMWH was significantly more effective than UH in preventing DVT in patients with acute ischemic strokes (13% vs 22%, odds ratio, 0.52) [196], and a subsequent review that included 11 trials involving 22,776 patients reported a decrease in pulmonary emboli (OR 0.60) [197]. However, these benefits were countered by increases in symptomatic intracranial and extracranial hemorrhages (odds ratios of 2.55 and 2.99, respectively), and there was no evidence that anticoagulant therapy within two weeks of stroke onset reduced the odds of death from all causes or dependency at the end of follow-up.

Summary

Table 4.3 lists the thrombotic disorders described in the text, and the heparins used in their management. In general, LMWHs are more effective and safer than UH and they can often be administered at home. Fondaparinux is at least as effective as LMWH, but its long half-life, exclusive renal excretion, and inability to prevent catheter thrombosis render it less suitable for

TABLE 4.3 Heparins in the management of thrombotic disorders.

Clinical disorder	Comment
Proximal DVT	LMWH more effective & safer than IV UH; alternatives are subcutaneous UH & fondaparinux
Pulmonary embolism	LMWH & fondaparinux are as effective & safe as IV UH
VTE in cancer	LMWH associated with better outcomes than vitamin K antagonists
Upper extremity DVT	LMWH effective & preserves catheter patency
Distal DVT	Heparins decrease recurrent VTE
Superficial VT	Fondaparinux prevents PE
Retinal VT	LMWH safe, but efficacy uncertain
Cerebral sinovenous thrombosis	LMWH associated with better outcomes; doses are reduced if major hemorrhages are present
Splanchnic VT	LMWH if thrombus of recent origin; adjust dose based on bleeding risk
Renal vein thrombosis	LMWH prevents recurrences; duration of therapy determined by ongoing risk factors
VT in pregnancy	LMWH safe & effective when dosed according to early pregnancy body weight.
VT in pediatrics	IV UH, with higher doses in infants & young children and monitoring of coagulation
Post-thrombotic syndrome	Heparin benefit appears limited to prevention of recurrent VT
Acute coronary syndromes	LMWH & fondaparinux more effective than UH, but more major bleeding with LMWH.
Percutaneous coronary intervention	LMWH effective but fondaparinux does not prevent catheter thrombosis
Mechanical heart valves; bridging	LMWH more effective than UH but more bleeding; in pregnancy, monitoring & dose-adjustments required.
Ischemic stroke	LMWH more effective than UH in preventing VTE, but both increase bleeding

Abbreviations: DVT = deep vein thrombosis; VTE = venous thromboembolism; LMWH = low molecular weight heparin; IV UH = intravenous unfractionated heparin.

use in pregnancy or percutaneous coronary interventions. All heparins contribute to bleeding, and might provoke hypersensitivity reactions, heparin-induced thrombocytopenia, and osteoporosis; these adverse effects are described in Chapter 5.

References

[1]   Marks J, Truscott BM, Withycombe JFR. Treatment of venous thrombosis with anticoagulants. The Lancet 1954;ii:787–91.

[2]   Barritt DW, Jordan SC. Anticoagulant drugs in the treatment of pulmonary embolism. The Lancet 1960;i:1309–12.

[3]   Gurewich V, Thomas DP, Stuart RK. Some guidelines for heparin therapy of venous thromboembolic disease. JAMA 1967;199:152–4.

[4]   Deykin D. The use of heparin. N Engl J Med 1969;280:937–8.

[5]   Basu D, Gallus A, Hirsh J, Cade J. A prospective study of value of monitoring heparin treatment with the activated partial thromboplastin time. N Engl J Med 1972;287:324–7.

[6]   Hattersley PG, Mitsuoka C, King JH. Sources of error in heparin therapy of thromboembolic disease. Arch Intern Med 1980;140:1173–5.

[7]   Chung KK, Tofferi JK, Browne WT. Suboptimal monitoring and dosing of unfractionated heparin. Ann Intern Med 2004;140:582.

[8]   Levine MN, Hirsh J, Gent M, Turpie AG, Cruickshank M, Weitz J, et al. A randomized trial comparing activated thromboplastin time with heparin assay in patients with acute venous thromboembolism requiring large daily doses of heparin. Arch Intern Med 1994;154:49–56.

[9]   Hull RD, Raskob GE, Brant RF, Pineo GF, Valentine KA. Relation between the time to achieve the lower limit of the APTT therapeutic range and recurrent venous thromboembolism during heparin treatment for deep vein thrombosis. Arch Intern Med 1997;157:2562–8.

[10]   Southwick F. Who was caring for Mary? Ann Intern Med 1993;118:146–8.

[11]   Hull RD, Raskob GE, Rosenbloom D, Lenaire J, Pineo GF, Baylis B, et al. Optimal therapeutic level of heparin therapy in patients with venous thrombosis. Arch Intern Med 1992;152:1589–95.

[12]   Raschke RA, Reilly BM, Guidry JR, Fontana JR, Srinivas S. The weight-based heparin dosing nomogram compared with a "standard care" nomogram. A randomized controlled trial. Ann Intern Med 1993;119:874–81.

[13]   Doyle DJ, Turpie AGG, Hirsh J, Best C, Kinch D, Levine MN, et al. Adjusted subcutaneous heparin or continuous intravenous heparin in patients with acute deep vein thromovsis. Ann Intern Med 1987;107:441–5.

[14]   Hommes DW, Bura A, Mazzolai L, Buller HR, ten Cate JW. Subcutaneous heparin compared with continuous intravenous heparin administration in the initial treatment of deep vein thrombosis: a meta-analysis. Ann Intern Med 1992;116:279–84.

[15]   Prandoni P, Bagatella P, Bernardi E, Girolami B, Rossi L, Scariano L, et al. Use of an algorithm for administering subcutaneous heparin in the treatment of deep venous thrombosis. Ann Intern Med 1998;129:299–302.

[16]   Hull RD, Raskob GE, Rosenbloom D, Panju AA, Brill-Edwards P, Ginsberg JS, et al. Heparin for 5 days as compared with 10 days in the initial treatment of proximal venous thrombosis. N Engl J Med 1990;322:1260–4.

[17]   Rooke TW, Osmundson PJ. Heparin and the in-hospital management of deep venous thrombosis: cost considerations. Mayo Clin Proc 1986;61:198–204.

[18]   Hull R, Delmore T, Carter C, Hirsh J, Genton E, Gent M, et al. Adjusted subcutaneous heparin versus warfarin sodium in the long-term treatment of venous thrombosis. N Engl J Med 1982;306:189–94.

[19] Hull RD, Raskob GE, Pineo GF, Green D, Trowbridge AA, Elliott CG, et al. Subcutaneous low-molecular-weight heparin compared with continuous intravenous heparin in the treatment of proximal-vein thrombosis. N Engl J Med 1992;326:975−82.

[20] Lindmarker P, Holmström M, Granqvist S, Johnsson H, Lockner D. Comparison of once-daily subcutaneous Fragmin with continuous intravenous unfractionated heparin in the treatment of deep vein thrombosis. Thromb Haemost 1994;72:186−90.

[21] Fiessinger JN, Lopez-Fernandez M, Gatterer E, Granqvist S, Kher A, Olsson CG, et al. Once-daily subcutaneous dalteparin, a low molecular weight heparin, for the initial treatment of acute deep vein thrombosis. Thromb Haemost 1996;76:195−9.

[22] Lee AY, Levine MN, Baker RI, Bowden C, Kakkar AK, Prins M, et al. Low-molecular-weight heparin versus a coumarin for the prevention of recurrent venous thromboembolism in patients with cancer. N Engl J Med 2003;349:146−53.

[23] Levine M, Gent M, Hirsh J, Leclerc J, Anderson D, Weitz J, et al. A comparison of low-molecular-weight heparin administered primarily at home with unfractionated heparin administered in the hospital for proximal deep-vein thrombosis. N Engl J Med 1996;334:677−81.

[24] Merli G, Spiro TE, Olsson C-G, Abildgaard U, Davidson BL, Eldor A, et al. Subcutaneous enoxaparin once or twice daily compared with intravenous unfractionated heparin for treatment of venous thromboembolic disease. Ann Intern Med 2001;134:191−202.

[25] Koopman MMW, Prandoni P, Piovella F, Ockelford PA, Brandjes DPM, van der Meer J, et al. Treatment of venous thrombosis with intravenous unfractionated heparin administered in the hospital as compared with subcutaneous low-molecular-weight heparin administered at home. N Engl J Med 1996;334:682−7.

[26] Wells PS, Kovacs MJ, Bormanis J, Forgie MA, Goudie D, Morrow B, et al. Expanding eligibility for outpatient treatment of deep venous thrombosis and pulmonary embolism with low-molecular-weight heparin: a comparison of patient self-injection with homecare injection. Arch Intern Med 1998;158:1809−12.

[27] Lensing AWA, Prins MH, Davidson BL, Hirsh J. Treatment of deep venous thrombosis with low-molecular-weight heparins. Arch Intern Med 1995;155:601−7.

[28] Gould MK, Dembitzer AD, Doyle RL, Hastie TJ, Garber AM. Low-molecular-weight heparins compared with unfractionated heparin for treatment of acute deep venous thrombosis: a meta-analysis of randomized, controlled trials. Ann Intern Med 1999;130:800−9.

[29] Hull RD, Raskob GE, Rosenbloom D, Pineo GF, Lerner RG, Gafni A, et al. Treatment of proximal vein thrombosis with subcutaneous low-molecular-weight heparin vs intravenous heparin. Arch Intern Med 1997;157:289−94.

[30] Gould MK, Dembitzer AD, Sanders GD, Garber AM. Low-molecular-weight heparins compared with unfractionated heparin for treatment of acute deep venous thrombosis. Ann Intern Med 1999;130:789−99.

[31] Kearon C, Ginberg JS, Julian JA, Douketis J, Solymoss S, Ockelford P, et al. A comparison of fixed-dose weight-adjusted unfractionated heparin and low-molecular-weight heparin for acute treatment of venous thromboembolism. JAMA 2006;296:935−42.

[32] Buller HR, Davidson BL, Decousus H, Gallus A, Gent M, Piovella F, et al. Fondaparinux or enoxaparin for the initial treatment of symptomatic deep venous thrombosis. A randomized trial. Ann Intern Med 2004;140:867−73.

[33] Partsch H, Kechavarz B, Kohn H, Mostbeck A. The effect of mobilization of patients during treatment of thromboembolic disorders with low-molecular-weight heparin. Int Angiol 1997;16:189−92.

[34] Aschwanden M, Labs K-H, Engel H, Schwob A, Jeanneret C, Mueller-Brand J, et al. Acute deep vein thrombosis: early mobilization does not increase the frequency of pulmonary embolism. Thromb Haemost 2001;85:42−6.

[35] Simonneau G, Sors H, Charbonnier B, Page Y, Laaban J-P, Azarian R, et al. A comparison of low-molecular-weight heparin with unfractionated heparin for acute pulmonary embolism. N Engl J Med 1997;337:663−9.

[36] Schutgens REG, Esseboom EU, Snijder RJ, Haas FJLM, Verzijlbergen F, Nieuwenhuis HK, et al. Low molecular weight heparin (dalteparin) is equally effective as unfractionated heparin in reducing coagulation activity and perfusion abnormalities during the early treatment of pulmonary embolism. J Lab Clin Med 2004;144:100−7.

[37] Quinlan DJ, McQuillan A, Eikelboom JW. Low-molecular-weight heparin compared with intravenous unfractionated heparin for treatment of pulmonary embolism: a meta-analysis of randomized, controlled trials. Ann Intern Med 2004;140:175−83.

[38] The Matisse Investigators. Subcutaneous fondaparinux versus intravenous unfractionated heparin in the initial treatment of pulmonary embolism. N Engl J Med 2003;349:1695−702.

[39] Kovacs MJ, Anderson D, Morrow B, Gray L, Touchie D, Wells PS. Outpatient treatment of pulmonary embolism with dalteparin. Thromb Haemost 2000;83:209−11.

[40] Aujesky D, Roy PM, Verschuren F, Righini M, Osterwalder J, Egloff M, et al. Outpatient versus inpatient treatment for patients with acute pulmonary embolism: an international, open-label, randomised, non-inferiority trial. The Lancet 2011;378:41−8.

[41] Zondag W, Vingerhoets LM, Durian MF, Dolsma A, Faber LM, Hiddinga BI, et al. Hestia criteria can safely select patients with pulmonary embolism for outpatient treatment irrespective of right ventricular function. J Thromb Haemost 2013;11:686−92.

[42] Kearon C, Akl EA. Duration of anticoagulant therapy for deep vein thrombosis and pulmonary embolism. Blood 2014;123:1794−801.

[43] Douketis JD, Gu CS, Schulman S, Ghiraduzzi A, Pengo V, Prandoni P. The risk for fatal pulmonary embolism after discontinuing anticoagulant therapy for venous thromboembolism. Ann Intern Med 2007;147:766−74.

[44] Monreal M, Lafoz E, Olive A, del Rio L, Vedia C. Comparison of subcutaneous unfractionated heparin with a low molecular weight heparin (Fragmin®) in patients with venous thromboembolism and contraindications to coumarin. Thromb Haemost 1994;71:7−11.

[45] Wilson SJ, Wilbur K, Burton E, Anderson DR. Effect of patient weight on the anticoagulant response to adjusted therapeutic dosage of low−molecular-weight heparin for the treatment of venous thromboembolism. Haemostasis 2001;31:42−8.

[46] Al-Yaseen E, Wells PS, Anderson J, Martin J, Kovacs MJ. The safety of dosing dalteparin based on actual body weight for the treatment of acute venous thromboembolism in obese patients. J Thromb Haemost 2005;3:100−2.

[47] Sanderink GJ, LeLiboux A, Jariwala N, Harding N, Ozoux ML, Shukla U, et al. The pharmacokinetics and pharmacodynamics of enoxaparin in obese volunteers. Clin Pharmacol Ther 2002;72:308−18.

[48] Lee YR, Vega JA, Duong HN, Ballew A. Monitoring enoxaparin with antifactor Xa levels in obese patients. Pharmacotherapy 2015;35:1007−15.

[49] Spinler SA, Inverso SM, Cohen M, Goodman SG, Stringer KA, Antman EM, et al. Safety and efficacy of unfractionated heparin versus enoxaparin in patients who are obese and patients with severe renal impairment: analysis from the ESSENCE and TIMI 11B studies. Am Heart J 2003;146:33−41.

[50]   Davidson BL, Buller HR, Decousus H, Gallus A, Gent M, Piovella F, et al. Matisse Investigators. Effect of obesity on outcomes after fondaparinux, enoxaparin, or heparin treatment for acute venous thromboembolism in the Matisse trials. J Thromb Haemost 2007;5:1191–4.

[51]   Hainer JW, Barrett JS, Assaid CA, Fossler MJ, Cox DS, Leathers T, et al. Dosing in heavy-weight/obese patients with the LMWH, tinzaparin: a pharmacodynamics study. Thromb Haemost 2002;87:817–23.

[52]   Witt DM, Nieuwlaat R, Clark NP, Ansell J, Holbrook A, Skov J, et al. American Society of Hematology 2018 guidelines for management of venous thromboembolism: optimal management of anticoagulation therapy. Blood Adv 2018;2:3257–91.

[53]   Douketis J, Cook D, Meade M, Guyatt G, Geerts W, Skrobik Y, et al. Prophylaxis against deep vein thrombosis in critically ill patients with severe renal insufficiency with the low-molecular-weight heparin dalteparin. Arch Intern Med 2008;168:1805–12.

[54]   Park D, Southern W, Calvo M, Kushnir M, Solorzano C, Sinnet M, et al. Treatment with dalteparin is associated with a lower risk of bleeding compared to treatment with unfractionated heparin in patients with renal insufficiency. J Gen Intern Med 2016;31:182–7.

[55]   Woodruff S, Feugere G, Abreu P, Heissler J, Ruiz MT, Jen F. A post hoc analysis of dalteparin versus oral anticoagulant (VKA) therapy for the prevention of recurrent venous thromboembolism (rVTE) in patients with cancer and renal impairment. J Thromb Thrombolysis 2016;42:494–504.

[56]   Siguret V, Pautas E, Fevrier M, Wipff C, Durand-Gasselin B, Laurent M, et al. Elderly patients treated with tinzaparin (Innohep®) administered once daily (175 anti-Xa IU/kg): anti-Xa and anti-IIa activities over 10 days. Thromb Haemost 2000;84:800–4.

[57]   Siguret V, Gouin-Thibault I, Pautas E, Leizorovicz A. No accumulation of the peak anti-factor Xa activity of tinzaparin in elderly patients with moderate-to-severe renal impairment: the IRIS substudy. J Thromb Haemost 2011;9:1966–72.

[58]   Leizorovicz A, Siguret V, Mottier D, Innohep® in Renal Insufficiency Study Steering Committee, Leizorovicz A, Siguret V, et al. Safety profile of tinzaparin versus subcutaneous unfractionated heparin in elderly patients with impaired renal function treated for acute deep vein thrombosis: the Innohep® in Renal Insufficiency Study (IRIS). Thromb Res 2011;128:27–34.

[59]   Lim W, Dentali F, Eikelboom JW, Crowther MA. Meta-analysis: low-molecular-weight heparin and bleeding in patients with severe renal insufficiency. Ann Intern Med 2006;144:673–84.

[60]   Mismetti P, Laporte-Simitsidis S, Navarro C, Sie P, d'Azemar P, Necciari J, et al. Aging and venous thromboembolism influence the pharmacodynamics of the anti-factor Xa and anti-thrombin activities of a low molecular weight heparin (nadroparin). Thromb Haemost 1998;79:1162–5.

[61]   Bauer KA. New pentasaccharides for prophylaxis of deep vein thrombosis: pharmacology. Chest 2003;124(6 Suppl):364S–70S.

[62]   Bauersachs RM, Lensing AW, Prins MH, Kubitza D, Pap ÁF, Decousus H, et al. Rivaroxaban versus enoxaparin/vitamin K antagonist therapy in patients with venous thromboembolism and renal impairment. Thromb J 2014;12:25.

[63]   Matzdorff A. Diagnosis and therapy in cancer-associated thromboembolism-what about guideline recommendations? Phlebologie 2015;44:299–303.

[64]   Hettiarachchi RJ, Prins MH, Lensing AW, Buller HR. Low molecular weight heparin versus unfractionated heparin in the initial treatment of venous thromboembolism. Curr Opin Pulm Med 1998;4:220–5.

[65] Morita S, Gebska MA, Kakkar AK, Scully MF. High affinity binding of heparin by necrotic tumour cells neutralizes anticoagulant activity. Thromb Haemost 2001;86:616–22.

[66] Akl EA, Kahale L, Barba M, Neumann I, Labedi N, Terrenato I, et al. Anticoagulation for the long-term treatment of venous thromboembolism in patients with cancer. Cochrane Database Syst Rev 2014;7:CD006650. Available from: https://doi.org/10.1002/14651858. CD006650.pub4.

[67] Prandoni P. How I treat venous thromboembolism in patients with cancer. Blood 2005;106:4027–33.

[68] Chai-Adisaksopha C, Iorio A, Crowther MA, de Miguel J, Salgado E, Zdraveska M, et al. Vitamin K antagonists after 6 months of low-molecular-weight heparin in cancer patients with venous thromboembolism. Am J Med 2018;131:430–7.

[69] Di Nisio M, Lee AY, Carrier M, Liebman HA, Khorana AA, for the Subcommittee on Haemostasis and Malignancy. Diagnosis and treatment of incidental venous thromboembolism in cancer patients: guidance from the SSC of the ISTH. J Thromb Haemost 2015;13:880–3.

[70] Kraaijpoel N, Bleker SM, Meyer G, Mahé I, Muñoz A, Bertoletti L, et al. Treatment and long-term clinical outcomes of incidental pulmonary embolism in patients with cancer: an international prospective cohort study. J Clin Oncol 2019;37:1713–20.

[71] Samuelson Bannow BT, Lee A, Khorana AA, Zwicker JI, Noble S, Ay C, et al. Management of cancer-associated thrombosis in patients with thrombocytopenia: guidance form the SSC of the ISTH. J Thromb Haemost 2018;16:1246–9.

[72] Schulman S, Zondag M, Linkins L, Pasca S, Cheung YW, De Sancho M, et al. Recurrent venous thromboembolism in anticoagulated patients with cancer: management and short-term prognosis. J Thromb Haemost 2015;13:1010–18.

[73] Kahale LA, Hakoum MB, Tsolakian IG, Matar CF, Terrenato I, Sperati F, et al. Anticoagulation for the long-term treatment of venous thromboembolism in people with cancer. Cochrane Database Syst Rev 2018;6:CD006650. Available from: https://doi.org/10.1002/14651858.CD006650.pub5.

[74] Kraaijpoel N, Di Nisio M, Mulder FI, van Es N, Beyer-Westendorf J, Carrier M, et al. Clinical impact of bleeding in cancer-associated venous thromboembolism: results from the Hokusai VTE cancer study. Thromb Haemost 2018;118:1439–49.

[75] Young AM, Marshall A, Thirlwall J, Chapman O, Lokare A, Hill C, et al. Comparison of an oral factor Xa inhibitor with low molecular weight heparin in patients with cancer with venous thromboembolism: results of a randomized trial (SELECT-D). J Clin Oncol 2018;36:2017–23.

[76] Khorana AA, Noble S, Lee YY, Soff G, Meyer G, O'Connell C, et al. Role of direct oral anticoagulants in the treatment of cancer-associated venous thromboembolism: guidance from the SSC of the ISTH. J Thromb Haemost 2018;16:1891–4.

[77] Kraaijpoel N, Carrier M. How I treat cancer-associated venous thromboembolism. Blood 2019;133:291–8.

[78] Francis CW, Kessler CM, Goldhaber SZ, Kovacs MJ, Monreal M, Huisman MV, et al. Treatment of venous thromboembolism in cancer patients with dalteparin for up to 12 months: the DALTECAN Study. J Thromb Haemost 2015;13:1028–35.

[79] Jara-Palomares L, Solier-Lopez A, Elias-Hernandez T, Asensio-Cruz M, Blasco-Esquivias I, Marin-Barrera L, et al. Tinzaparin in cancer associated thrombosis beyond 6months: TiCAT study. Thromb Res 2017;157:90–6.

[80] Savage KJ, Wells PS, Schulz V, Goudie D, Morrow B, Cruickshank M, et al. Outpatient use of low molecular weight heparin (dalteparin) for the treatment of deep vein thrombosis of the upper extremity. Thromb Haemost 1999;82:1008−10.

[81] Kovacs MJ, Kahn SR, Rodger M, Anderson DR, Andreou R, Mangel JE, et al. A pilot study of central venous catheter survival in cancer patients using low-molecular-weight heparin (dalteparin) and warfarin without catheter removal for the treatment of upper extremity deep vein thrombosis (The Catheter Study). J Thromb Haemost 2007;5:1650−3.

[82] Rathbun SW, Stoner JA, Whitsett TL. Treatment of upper-extremity deep vein thrombosis. J Thromb Haemost 2011;9:1924−30.

[83] Zwicker JI, Connolly G, Carrier M, Kamphuisen PW, Lee AYY. Catheter-associated deep vein thrombosis of the upper extremity in cancer patients: guidance from the SSC of the ISTH. J Thromb Haemost 2014;12:796−800.

[84] Davies GA, Lazo-Langner A, Gandara E, Rodger M, Tagalakis V, Louzada M, et al. A prospective study of Rivaroxaban for central venous catheter associated upper extremity deep vein thrombosis in cancer patients (Catheter 2). Thromb Res 2018;162:88−92.

[85] Montiel FS, Ghazvinian R, Gottsater A, Elf J. Treatment with direct oral anticoagulants in patients with upper extremity deep vein thrombosis. Thromb J 2017;15:26. Available from: https://doi.org/10.1186/s12959-017-0149-x.

[86] Bentley PG, Kakkar VV, Scully MF, MacGregor IR, Webb P, Chan P, et al. An objective study of alternative methods of heparin administration. Thromb Res 1980;18:177−87.

[87] Lagerstedt CI, Olsson C-G, Fagher BO, Oqvist BW, Albrechtsson U. Need for long-term anticoagulant treatment in symptomatic calf-vein thrombosis. The Lancet 1985;ii:515−18.

[88] Schwarz T, Buschmann L, Beyer J, Halbritter K, Rastan A, Schellong S. Therapy of isolated calf muscle vein thrombosis: a randomized, controlled study. J Vasc Surg 2010;52:1246−50.

[89] Righini M, Galanaud JP, Guenneguez H, Brisot D, Diard A, Faisse P, et al. Anticoagulant therapy for symptomatic calf deep vein thrombosis (CACTUS): a randomised, double-blind, placebo-controlled trial. Lancet Haematol 2016;3:e556−62.

[90] Franco L, Giustozzi M, Agnelli G, Becattini C. Anticoagulation in patients with isolated distal deep vein thrombosis: a meta-analysis. J Thromb Haemost 2017;15:1142−54.

[91] Kearon C, Akl EA, Ornelas J, Blaivas A, Jimenez D, Bounameaux H, et al. Antithrombotic Therapy for VTE disease: CHEST guideline and expert panel report. Chest 2016;149:315−52.

[92] The Vesalio Investigators Group. High vs low doses of low-molecular-weight heparin for the treatment of superficial vein thrombosis of the legs: a double-blind, randomized trial. J Thromb Haemost 2005;3:1152−7.

[93] Rathbun SW, Aston CE, Whitsett TL. A randomized trial of dalteparin compared with ibuprofen for the treatment superficial thrombophlebitis. J Thromb Haemost 2012;10:833−9.

[94] Decousus H, Prandoni P, Mismetti P, Bauersachs RM, Boda Z, Brenner B, et al. Fondaparinux for the treatment of superficial-vein thrombosis in the legs. N Engl J Med 2010;363:1222−32.

[95] Di Nisio M, Wichers IM, Middeldorp S. Treatment for superficial thrombophlebitis of the leg. Cochrane Database Syst Rev 2012;CD004982.

[96] Beyer-Westendorf J, Schellong SM, Gerlach H, Rabe E, Weitz JI, Jersemann K, et al. Prevention of thromboembolic complications in patients with superficial-vein thrombosis given rivaroxaban or fondaparinux: the open-label, randomised, non-inferiority SURPRISE phase 3b trial. Lancet Haematol 2017;4:e105−13.

[97]   Lazo-Langner A, Hawel J, Ageno W, Kovacs MJ. Low molecular weight heparin for the treatment of retinal vein occlusion: a systematic review and meta-analysis of randomized trials. Haematologica 2010;95:1587—93.

[98]   Wong TY, Scott IU. Retinal-vein occlusion. N Engl J Med 2011;364:979.

[99]   Sartori MT, Barbar S, Dona A, Piermarocchi S, Pilotto E, Saggiorato G, et al. Risk factors, antithrombotic treatment and outcome in retinal vein occlusion: an age-related prospective cohort study. Eur J Haematol 2013;90:426—33.

[100] Steigerwalt Jr RD, Pascarella A, DeAngelis M, Ciucci F, Gaudenzi F. An ultra-low-molecular-weight heparin, fondaparinux, to treat retinal vein occlusion. Drug Discov Ther 2016;10:167—71.

[101] Bousser M-G. Cerebral venous thrombosis: nothing, heparin, or local thrombolysis? Stroke 1999;30:481—3.

[102] Einhaupl KM, Villringer A, Meister W, Mehraein S, Garner C, Pellkofer M, et al. Heparin treatment in sinus venous thrombosis. The Lancet 1991;338:597—600.

[103] De Bruijn SF, Stam J. Randomized, placebo-controlled trial of anticoagulant treatment with low-molecular-weight heparin for cerebral sinus thrombosis. Stroke 1999;30:484—8.

[104] Coutinho JM, Ferro JM, Canhao P, Barinagarrementeria F, Bousser MG, Stam J. ISCVT investigators. Unfractionated or low-molecular weight heparin for the treatment of cerebral venous thrombosis. Stroke 2010;41:2575—80.

[105] Misra UK, Kalita J, Chandra S, Kumar B, Bansal V. Low molecular weight heparin versus unfractionated heparin in cerebral venous sinus thrombosis: a randomized controlled trial. Eur J Neurol 2012;19:1030—6.

[106] Afshari D, Moradian N, Nasiri F, Razazian N, Bostani A, Sariaslani P. The efficacy and safety of low-molecular-weight heparin and unfractionated heparin in the treatment of cerebral venous sinus thrombosis. Neurosciences (Riyadh) 2015;20:357—61.

[107] Ferro JM, Bousser MG, Canhão P, Coutinho JM, Crassard I, Dentali F, et al. European Stroke Organization guideline for the diagnosis and treatment of cerebral venous thrombosis - endorsed by the European Academy of Neurology. Eur J Neurol 2017;24:1203—13.

[108] Capecchi M, Abbattista M, Martinelli I. Cerebral venous sinus thrombosis. J Thromb Haemost 2018;16:1918—31.

[109] Sogaard KK, Farkas DK, Pedersen L, Sorensen HT. Splanchnic venous thrombosis is a marker of cancer and a prognostic factor for cancer survival. Blood 2015;126:957—63.

[110] Kearon C, Akl EA, Comerota AJ, Prandoni P, Bounameaux H, Goldhaber SZ, et al. Antithrombotic therapy for VTE disease: antithrombotic therapy and prevention of thrombosis, 9th ed: American College of Chest Physicians Evidence-Based Clinical Practice Guidelines. Chest 2012;141(2 Suppl):e419S—96S.

[111] Riva N, Ageno W, Schulman S, Beyer-Westendorf J, Duce R, Malato A, et al. Clinical history and antithrombotic treatment of incidentally detected splanchnic vein thrombosis: a multicenter, international prospective registry. Lancet Haematol 2016;3:e267—75.

[112] Ageno W, Dentali F, Squizzato A. How I treat splanchnic vein thrombosis. Blood 2014;124:3685—91.

[113] Kwon J, Koh Y, Yu SJ, Yoon JH. Low-molecular-weight heparin treatment for portal vein thrombosis in liver cirrhosis: efficacy and the risk of hemorrhagic complications. Thromb Res 2018;163:71—6.

[114]   Senzolo M, Riva N, Dentali F, Rodriguez-Castro K, Sartori MT, Bang SM, et al. Long-term outcome of splanchnic vein thrombosis in cirrhosis. Clin Transl Gastroenterol 2018;9:176.

[115]   Loscalzo J. Venous thrombosis in the nephrotic syndrome. N Engl J Med 2013;368:956−8.

[116]   De Stefano V, Martinelli I. Abdominal thrombosis of splanchnic, renal and ovarian veins. Best Pract Res Clin Haematol 2012;25:253−64.

[117]   Obernosterer A, Aschauer M, Schnedl W, Lipp RW. Anomalies of the inferior vena cava in patients with iliac venous thrombosis. Ann Intern Med 2002;136:37−41.

[118]   Greer IA. Pregnancy complicated by venous thrombosis. N Engl J Med 2015;373:540−7.

[119]   Blondon M, Casini A, Hoppe KK, Boehlen F, Righini M, Smith NL. Risks of venous thromboembolism after cesarean sections: a meta-analysis. Chest 2016;150:572−96.

[120]   James AH, Jamison MG, Brancazio LR, Myers ER. Venous thromboembolism during pregnancy and the postpartum period: incidence, risk factors, and mortality. Am J Obstet Gynecol 2006;194:1311−15.

[121]   Bates SM, Ginsberg JS. How we manage venous thromboembolism during pregnancy. Blood 2002;100:3470−8.

[122]   Romualdi E, Dentali F, Rancan E, Squizzato A, Steidl L, Middeldorp S, et al. Anticoagulant therapy for venous thromboembolism during pregnancy: a systematic review and a meta-analysis of the literature. J Thromb Haemost 2013;11:270−81.

[123]   Marik PE, Plante LA. Venous thromboembolic disease and pregnancy. N Engl J Med 2008;359:2025−33.

[124]   Bates SM, Rajasekhar A, Middeldorp S, McLintock C, Rodger MA, James AH, et al. American Society of Hematology 2018 guidelines for management of venous thrombo-embolism: venous thromboembolism in the context of pregnancy. Blood Adv 2018;2:3317−59.

[125]   Horlocker TT, Wedel DJ, Benzon H, Brown DL, Enneking FK, Heit JA, et al. Regional anesthesia in the anticoagulated patient: defining the risks (the second ASRA Consensus Conference on Neuraxial Anesthesia and Anticoagulation). Reg Anesth Pain Med 2003;28:172−97.

[126]   Monagle P, Newall F. Management of thrombosis in children and neonates: practical use of anticoagulants in children. Hematology 2018.

[127]   Hepponstall M, Chan A, Monagle P. Anticoagulation therapy in neonates, children and adolescents. Blood Cells Mol Dis 2017;67:41−7.

[128]   Monagle P, Cuello CA, Augustine C, Bonduel M, Brandão LR, Capman T, et al. American Society of Hematology 2018 Guidelines for management of venous thrombo-embolism: treatment of pediatric venous thromboembolism. Blood Adv 2018;2:3292−316.

[129]   Baldwin MJ, Moore HM, Rudarakanchana N, Gohel M, Davies AH. Post-thrombotic syndrome: a clinical review. J Thromb Haemost 2013;11:795−805.

[130]   Rabinovich A, Kahn SR. How I treat the postthrombotic syndrome. Blood 2018;131:2215−22.

[131]   van Dongen CJ, Prandoni P, Frulla M, Marchiori A, Prins MH, Hutten BA. Relation between quality of anticoagulant treatment and the development of the postthrombotic syndrome. J Thromb Haemost 2005;3:939−42.

[132]   González-Fajardo JA, Martin-Pedrosa M, Castrodeza J, Tamames S, Vaquero-Puerta C. Effect of the anticoagulant therapy in the incidence of post-thrombotic syndrome and

recurrent thromboembolism: comparative study of enoxaparin versus coumarin. J Vasc Surg 2008;48:953−9.

[133] Hull RD, Pineo GF, Brant R, Liang J, Cook R, Solymoss S, et al. Home therapy of venous thrombosis with long-term LMWH versus usual care: patient satisfaction and post-thrombotic syndrome. Am J Med 2009;122:762−9.

[134] Hull RD, Liang J, Townshend G. Long-term low-molecular-weight heparin and the post-thrombotic syndrome: a systematic review. Am J Med 2011;124:756−65.

[135] San Norberto EM, Gastambide MV, Taylor JH, García-Saiz I, Vaquero C. Effects of rosuvastatin as an adjuvant treatment for deep vein thrombosis. Vasa 2016;45:133−40.

[136] Cheung YW, Middeldorp S, Prins MH, Pap AF, Lensing AW, ten Cate-Hoek AJ, et al. Post-thrombotic syndrome in patients treated with rivaroxaban or enoxaparin/vitamin K antagonists for acute deep-vein thrombosis. A post-hoc analysis. Thromb Haemost 2016;116:733−8.

[137] Jeraj L, Jezovnik MK, Poredos P. Rivaroxaban versus warfarin in the prevention of post-thrombotic syndrome. Thromb Res 2017;157:46−8.

[138] Sogaard M, Nielsen PB, Skioth F, Kjaeldgaard JN, Coleman CI, Larsen TB. Rivaroxaban versus warfarin and risk of post-thrombotic syndrome among patients with venous thromboembolism. Am J Med 2018;131:787−94.

[139] Stain M, Schonauer V, Minar E, Bialonczyk C, Hirschl M, Weltermann A, et al. The post-thrombotic syndrome: risk factors and impact on the course of thrombotic disease. J Thromb Haemost 2005;3:2671−6.

[140] Vedantham S, Goldhaber SZ, Julian JA, Kahn SR, Jaff MR, Cohen DJ, et al. Pharmacomechanical catheter-directed thrombolysis for deep-vein thrombosis. N Engl J Med 2017;377:2240−52.

[141] Heras M, Chesebro JH, Webster WI, Mruk JS, Grill DE, Fuster V. Antithrombotic efficacy of low-molecular-weight heparin in deep arterial injury. Arterioscler Thromb 1992;12:250−5.

[142] Currier JW, Pow TK, Haudenschild CC, Minihan AC, Faxon DP. Low molecular weight heparin (enoxaparin) reduces restenosis after iliac angioplasty in the hypercholesterolemic rabbit. J Am Coll Cardiol 1991;17:118B−125BB.

[143] Turpie AGG, Robinson JG, Doyle DJ, Mulji AS, Mishkel GJ, Sealey BJ, et al. Comparison of high-dose with low-dose subcutaneous heparin to prevent left ventricular mural thrombosis in patients with acute transmural anterior myocardial infarction. N Engl J Med 1989;320:325−57.

[144] Cairns JA, Gent M, Singer J, Finnie KJ, Froggatt GM, Holder DA, et al. Aspirin, sulfinpyrazone, or both in unstable angina. Results of a Canadian multicenter trial. N Engl J Med 1985;313:1369−75.

[145] Theroux P, Ouimet H, McCans J, Latour J-G, Joly P, Levy G, et al. Aspirin, heparin, or both to treat acute unstable angina. N Engl J Med 1988;319:1105−11.

[146] Oler A, Whooley MA, Oler J, Grady D. Adding heparin to aspirin reduces the incidence of myocardial infarction and death in patients with unstable angina. A meta-analysis. JAMA 1996;276:811−15.

[147] Collins R, Peto R, Baigent C, Sleight P. Aspirin, heparin, and fibrinolytic therapy in suspected acute myocardial infarction. N Engl J Med 1997;336:847−60.

[148] Gurfinkel EP, Manos EJ, Mejaíl RI, Cerdá MA, Duronto EA, García CN, et al. Low molecular weight heparin versus regular heparin or aspirin in the treatment of unstable angina and silent ischemia. J Am Coll Cardiol 1995;26:313 18.

[149] Fragmin during instability in Coronary Artery Disease (FRISC) study group. Low-molecular-weight heparin during instability in coronary artery disease. The Lancet 1996;347:561−8.

[150] Ernofsson M, Strekerud F, Toss H, Abildgaard U, Wallentin L, Siegbahn A. Low-molecular weight heparin reduces the generation and activity of thrombin in unstable coronary artery disease. Thromb Haemost 1998;79:491−4.

[151] Klein W, Buchwald A, Hillis SE, Monrad S, Sanz G, Turpie AGG, et al. Comparison of low-molecular-weight heparin with unfractionated heparin acutely and with placebo for 6 weeks in the management of unstable coronary artery disease. Circulation 1997;96:61−8.

[152] FRagmin and Fast Revascularisation during InStability in Coronary artery disease (FRISC II) Investigators. Long-term low-molecular-mass heparin in unstable coronary-artery disease: FRISC II prospective randomized multicenter study. The Lancet 1999;354:701−7.

[153] Cohen M, Demers C, Gurfinkel EP, Turpie AGG, Fromell GJ, Goodman S, et al. A comparison of low−molecular-weight heparin with unfractionated heparin for unstable coronary artery disease. N Engl J Med 1997;337:447−52.

[154] Goodman SG, Cohen M, Bigonzi F, Gurfinkel EP, Radley DR, Le Iouer V, et al. Randomized trial of low molecular weight heparin (enoxaparin) versus unfractionated heparin for unstable coronary artery disease: one-year results of the ESSENCE Study. Efficacy and safety of subcutaneous enoxaparin in non-Q-wave coronary events. J Am Coll Cardiol 2000;36:693−8.

[155] Montalescot G, Philippe F, Ankri A, Vicaut E, Bearez E, Poulard JE, et al. Early increase of von Willebrand factor predicts adverse outcome in unstable coronary artery disease: beneficial effects of enoxaparin. Circulation 1998;98:294−9.

[156] Mark DB, Cowper PA, Berkowitz SD, Davidson-Ray L, DeLong ER, Turpie AGG, et al. Economic assessment of low-molecular-weight heparin (enoxaparin) versus unfractionated heparin in acute coronary syndrome patients: results from the ESSENCE randomized trial. Circulation 1998;97:1702−7.

[157] Antman EM, McCabe CH, Gurfinkel EP, Turpie AG, Bernink PJ, Salein D, et al. Enoxaparin prevents death and cardiac ischemic events in unstable angina/non-Q-wave myocardial infarction. Results of the thrombolysis in myocardial infarction (TIMI) 11B trial. Circulation 1999;100:1593−601.

[158] Goodman SG, Fitchett D, Armstrong PW, Tan M, Langer A, Integrilin and Enoxaparin Randomized Assessment of Acute Coronary Syndrome Treatment (INTERACT) Trial Investigators. Randomized evaluation of the safety and efficacy of enoxaparin versus unfractionated heparin in high-risk patients with non-ST-segment elevation acute coronary syndromes receiving the glycoprotein IIb/IIIa inhibitor eptifibatide. Circulation 2003;107:238−44.

[159] The Assessment of the Safety and Efficacy of a New Thrombolytic Regimen (ASSENT)-3 Investigators. Efficacy and safety of tenecteplase in combination with enoxaparin, abciximab, or unfractionated heparin: the ASSENT-3 randomised trial in acute myocardial infarction. The Lancet 2001;358:605−13.

[160] Antman EM, Morrow DA, McCabe CH, Murphy SA, Ruda M, Sadowski Z, et al. Enoxaparin versus unfractionated heparin with fibrinolysis for ST-elevation myocardial infarction. N Engl J Med 2006;354:1477−88.

[161] Eikelboom JW, Anand SS, Malmberg K, Weitz JI, Ginsberg JS, Yusuf S. Unfractionated heparin and low −molecular-weight heparin in acute coronary syndrome without ST elevation: a meta-analysis. The Lancet 2000;355:1936−42.

[162] Murphy SA, Gibson CM, Morrow DA, van de Werf F, Menown IB, Goodman SB, et al. Efficacy and safety of the low-molecular-weight heparin enoxaparin compared with unfractionated heparin across the acute coronary syndrome spectrum: a meta-analysis. Eur Heart J 2007;28:2077−86.

[163] Yusuf S, Mehta SR, Xie C, Ahmed RJ, Xavier D, Pais P, et al. Reviparin reduced a composite endpoint of death, reinfarction, stroke, and ischemia at 7 and 30 days after acute MI. JAMA 2005;293:427−36.

[164] Simoons ML, Bobbink IW, Boland J, Gardien M, Klootwijk P, Lensing AW, et al. A dose-finding study of fondaparinux in patients with non-ST-segment elevation acute coronary syndromes: the Pentasaccharide in Unstable Angina (PENTUA) Study. J Am Coll Cardiol 2004;43:2183−90.

[165] The Fifth Organization to Assess Strategies in Acute Ischemic Syndromes Investigators. Comparison of fondaparinux and enoxaparin in acute coronary syndromes. N Engl J Med 2006;354:1464−76.

[166] Anderson JA, Hirsh J, Yusuf S, Johnston M, Afzal R, Mehta SR, et al. Comparison of the anticoagulant intensities of fondaparinux and enoxaparin in the Organization to Assess Strategies in Acute Ischemic Syndromes (OASIS)-5 trial. J Thromb Haemost 2010;8:243−9.

[167] Budaj A, Eikelboom JW, Mehta SR, Afzal R, Chrolavicius S, Bassand JP, et al. Improving clinical outcomes by reducing bleeding in patients with non-ST-elevation acute coronary syndromes. Eur Heart J 2009;30:655−61.

[168] Yusuf S, Mehta SR, Chrolavicius S, Afzal R, Pogue J, Granger CB, et al. Effects of fondaparinux on mortality and reinfarction in patients with acute ST-segment elevation myocardial infarction: the OASIS-6 randomized trial. JAMA 2006;295:1519−30.

[169] Coussement PK, Bassand JP, Convens C, Vrolix M, Boland J, Grollier G, et al. A synthetic factor-Xa inhibitor (ORG31540/SR9017A) as an adjunct to fibrinolysis in acute myocardial infarction. The PENTALYSE study. Eur Heart J 2001;22:1716−24.

[170] Mega JL, Braunwald E, Wiviott SD, Bassand JP, Bhatt DL, Bode C, et al. Rivaroxaban in patients with a recent acute coronary syndrome. N Engl J Med 2012;366:9−19.

[171] Eikelboom JW, Connolly SJ, Bosch J, Dagenais GR, Hart RG, Shestakovska O, et al. Rivaroxaban with or without aspirin in stable cardiovascular disease. N Engl J Med 2017;377:1319−30.

[172] Chiarito M, Cao D, Cannata F, Godino C, Lodigiani C, Ferrante G, et al. Direct oral anticoagulants in addition to antiplatelet therapy for secondary prevention after acute coronary syndromes: a systematic review and meta-analysis. JAMA Cardiol 2018;3:234−41.

[173] Montalescot G, White HD, Gallo R, Cohen M, Steg PG, Aylward PEG, et al. Enoxaparin versus unfractionated heparin in elective percutaneous coronary intervention. N Engl J Med 2006;355:1006−17.

[174] Brieger D, Collet JP, Silvain J, Landivier A, Barthelemy O, Beygui F, et al. Heparin or enoxaparin anticoagulation for primary percutaneous coronary intervention. Catheter Cardiovasc Interv 2011;77:182−90.

[175] He P, Liu Y, Wei X, Jiang L, Guo W, Guo Z, et al. Comparison of enoxaparin and unfractionated heparin in patients with non-ST-elevation acute coronary syndrome

undergoing percutaneous coronary intervention: a systematic review and meta-analysis. J Thorac Dis 2018;10:3308−18.

[176] Vuillemenot A, Schiele F, Meneveau N, Claudel S, Donat F, Fontecave S, et al. Efficacy of a synthetic pentasaccharide, a pure factor Xa inhibitor, as an antithrombotic agent-a pilot study in the setting of coronary angioplasty. Thromb Haemost 1999;81:214−20.

[177] Mehta SR, Steg PG, Granger CB, Bassand J-P, Faxon DP, Weitz JI, et al. Randomized, blinded trial comparing fondaparinux with unfractionated heparin in patients undergoing contemporary percutaneous coronary intervention. Circulation 2005;111:1390−7.

[178] Yau JW, Stafford AR, Liao P, Fredenburgh JC, Roberts R, Weitz JI. Mechanism of catheter thrombosis: comparison of the antithrombotic activities of fondaparinux, enoxaparin, and heparin in vitro and in vivo. Blood 2011;118:6667−74.

[179] Douketis JD. Anticoagulation therapy in patients with prosthetic heart valves. Clin Adv Hematol Oncol. 2007;5:436−8.

[180] Kindo M, Gerelli S, Hoang Minh T, Zhang M, Meyer N, Announe T, et al. Exclusive low-molecular-weight heparin as bridging anticoagulant after mechanical valve replacement. Ann Thorac Surg 2014;97:789−95.

[181] Passaglia LG, de Barros GM, de Sousa MR. Early postoperative bridging anticoagulation after mechanical heart valve replacement: a systematic review and meta-analysis. J Thromb Haemost 2015;13:1557−67.

[182] Douketis JD, Spyropoulos AC, Spencer FA, Mayr M, Jaffer AK, Eckman MH, et al. Perioperative management of antithrombotic therapy: antithrombotic therapy and prevention of thrombosis, 9th ed: American College of Chest Physicians Evidence-Based Clinical Practice Guidelines. Chest 2012;141(2 Suppl):e326S−50S.

[183] Lim WY, Lloyd G, Bhattacharyya S. Mechanical and surgical bioprosthetic valve thrombosis. Heart 2017;103:1934−41.

[184] Clark NP, Douketis JD, Hasselblad V, Schulman S, Kindzelski AL, Ortel TL, et al. Predictors of perioperative major bleeding in patients who interrupt warfarin for an elective surgery or procedure: analysis of the BRIDGE trial. Am Heart J 2018;195:108−14.

[185] O'Donnell MJ, Kearon C, Johnson J, Robinson M, Zondag M, Turpie I, et al. Brief communication: preoperative anticoagulant activity after bridging low-molecular-weight heparin for temporary interruption of warfarin. Ann Intern Med 2007;146:184−7.

[186] Dhillon SK, Edwards J, Wilkie J, Bungard TJ. High-versus low-dose warfarin-related teratogenicity: a case report and systematic review. J Obstet Gynaecol Can 2018;40:1348−57.

[187] Goland S, Schwartzenberg S, Fan J, Kozak N, Khatri N, Elkayam U. Monitoring of anti-Xa in pregnant patients with mechanical prosthetic valves receiving low-molecular-weight heparin: peak or trough levels? J Cardiovasc Pharmacol Ther 2014;19:451−6.

[188] Vause S, Clarke B, Tower CL, Hay C, Knight M, (on behalf of UKOSS). Pregnancy outcomes in women with mechanical prosthetic heart valves: a prospective descriptive population based study using the United Kingdom Obstetric Surveillance System (UKOSS) data collection system. BJOG 2017;124:1411−19.

[189] Bates SM, Greer IA, Middeldorp S, Veenstra DL, Prabulos A-M, Vandvik PO. VTE, thrombophilia, antithrombotic therapy, and pregnancy. Chest 2012;141(2 Suppl): e691S−736S.

[190] Putman SF, Adams Jr. HP. Usefulness of heparin in initial management of patients with recent transient ischemic attacks. Arch Neurol 1985;42:960−2.

[191] Duke RJ, Bloch RF, Turpie AGG, Trebilcock R, Bayer N. Intravenous heparin for the prevention of stroke progression in acute partial stable stroke: a randomized controlled trial. Ann Intern Med 1986;105:825–8.

[192] Keith DS, Phillips SJ, Whisnant JP, Nishimaru K, O'Fallon WM. Heparin therapy for recent transient focal cerebral ischemia. Mayo Clin Proc 1987;62:1101–6.

[193] Kay R, Wong KS, Yu YL, Chan YW, Tsoi TH, Ahuja AT, et al. Low-molecular-weight heparin for the treatment of acute ischemic stroke. N Engl J Med 1995;333:1588–93.

[194] Berge E, Abdelnoor M, Nakstad PH, Sandset PM. Low molecular-weight heparin versus aspirin in patients with acute ischaemic stroke and atrial fibrillation: a double-blind randomised study. HAEST Study Group. Heparin in Acute Embolic Stroke Trial. The Lancet 2000;355:1205–10.

[195] Bath PMW, Lindenstrom E, Boysen G, De Deyn P, Friis P, Leys D, et al. Tinzaparin in acute ischaemic stroke (TAIST): a randomized aspirin-controlled trial. The Lancet 2001;358:702–10.

[196] Counsell C, Sandercock P. Low-molecular-weight heparins or heparinoids versus standard unfractionated heparin for acute ischemic stroke (Cochrane Review). Stroke 2002;33:1925–6.

[197] Sandercock PA, Counsell C, Kane EJ. Anticoagulants for acute ischaemic stroke. Cochrane Database Sys Rev 2016;12:CD000024.

Part III

Adverse Effects

Chapter 5

Bleeding, Heparin-induced thrombocytopenia, and Other adverse reactions

The principle adverse effects of heparin are bleeding, heparin-induced thrombocytopenia (HIT), allergic reactions, and osteoporosis. In addition, hyperkalemia has been reported infrequently and serum transaminases often rise during therapy. This Chapter focuses on bleeding and HIT, and provides brief descriptions of a few of the other side effects of heparin therapy.

Bleeding

The factors that increase the risk of bleeding with heparins are shown in Table 5.1. They include poor performance status, a history of a bleeding tendency, recent trauma or surgery, a body surface area <2 m^2, and exposure to high doses of heparins [1]. An analysis of data from 6 studies showed a direct correlation between the major bleeding rate and the total daily dose of heparin ($r = 0.73$, $P < 0.001$) [2]. Other risk factors are older age, a history of gastrointestinal or genitourinary bleeding, hepatic or renal impairment, cancer, thrombocytopenia, anemia, and concomitant use of antiplatelet or fibrinolytic agents [3]. The estimated absolute risk of bleeding is 1.6% with no risk factors, 3.2% with 1 risk factor, and 12.8% with ≥ 2 risk factors [3]. Although anticoagulant-associated bleeding is more likely in patients with cancer, the clinical course is not more severe than in non-cancer patients, and anticoagulation does not appear to increase the risk for intracranial hemorrhage in patients with cerebral metastasis [4,5].

It was recognized as early as 1968 that heparin provoked hemorrhage more often in women than men (32% versus 15%, $P < 0.003$), and particularly in women over age 6o (50% versus 14%, $P < 0.01$) [6]. Bleeding was also more likely in patients with lower body weights [7]. Common sites of bleeding were the gastrointestinal tract (9%), skin and muscle (7%), and the genitourinary tract (4%). Another study found that heparin levels were increased in the elderly, whose heparin dose requirements were lower after therapeutic activated partial thromboplastin times (aPTT) were achieved [8]. It was concluded

The Heparins. DOI: https://doi.org/10.1016/B978-0-12-818781-4.00005-4

TABLE 5.1 Factors associated with an increased risk of bleeding in patients receiving heparins.

Referable to the patient:

WHO[a] performance status grades 3 and 4 (non-ambulatory, severely impaired)

History of gastrointestinal or genitourinary bleeding

Recent trauma or surgery

Hepatic impairment

Renal failure

Cancer

Thrombocytopenia or bleeding tendency

Older age

Body surface area <2 m^2

Concomitant therapy with antiplatelet or fibrinolytic agents

Referable to the anticoagulant:

Higher doses and longer duration of exposure

Unfractionated heparin compared to low molecular weight heparin and fondaparinux

[a]*WHO = World Health Organization.*

that the pharmacokinetics of heparin are altered by aging, and probably explained the heightened sensitivity to heparin in elderly women [9].

An epidemiological study conducted in 1978 of 16,646 medical inpatients with no known predisposing illness found that gastrointestinal bleeding that was major in 1.2% and minor in 8.3% occurred after exposure to heparin [10]. A small study of 76 patients reported bleeding in 32% that was retroperitoneal in 7%, but hemorrhages were uncommon during the first 2 days of heparin exposure [11]. By the early 1990s, clinical trials had identified effective and relatively safe treatment doses of unfractionated heparin (UH). Major bleeding was defined as overt and associated with a fall in hemoglobin of 2 g/dL or more, led to transfusion of 2 or more units of blood, or was retroperitoneal, into a prosthetic joint, or intracranial. The incidence of major bleeding in patients receiving intravenous UH was reported to be 5% (11 of 219) [12]. A meta-analysis noted similar bleeding rates with subcutaneous and intravenous UH (4.1% and 5.2%) [13]. The rates of major bleeding have been lower in subsequent studies of intravenous UH; a meta-analysis of 14 trials of patients receiving treatment for pulmonary embolism noted a bleeding rate of 2.3% [14]. The use of UH has been associated with more bleeding events in patients having heart valve surgery (10%), and these events included hemothorax as well as pericardial and retroperitoneal bleeding [15].

TABLE 5.2 Major bleeding (%) with heparins used for the prevention and treatment of venous thrombosis.[a]

Product	Prevention	Treatment
Unfractionated Heparin	3.5 (2.6−6.0)	2.3
Low Molecular Weight Heparin		
Dalteparin	1.5	2.0
Enoxaparin	1.7	2.1
Tinzaparin	2.0	1.5
Synthetic Heparin (fondaparinux)	2.7	1.2

[a]Crowther MA, Warkentin TE. Bleeding risk and the management of bleeding complications in patients undergoing anticoagulant therapy: focus on new anticoagulant agents. Blood 2008;111:4871-79. Laporte S, Bertoletti L, Romera A, Mismetti P, Pérez de Llano LA, Meyer G. Long-term treatment of venous thromboembolism with tinzaparin compared to vitamin K antagonists: a meta-analysis of 5 randomized trials in non-cancer and cancer patients. Thromb Res 2012;130:853-58.

Major bleeding rates with heparins used for the prophylaxis and treatment of venous thrombosis are displayed in Table 5.2. When these anticoagulants are used for the treatment of thrombotic disorders, major bleeding is equivalent or less with low molecular weight heparin (LMWH) than with intravenous UH. A meta-analysis of 11 studies including 3674 patients reported that the rate of major bleeding was lower with LMWH than intravenous UH (1.1% vs 1.9%, P = 0.047) although the absolute risk reduction was small and not statistically significant [16]. On the other hand, a study that compared LMWH with subcutaneous UH found similar rates of major bleeding (1.4% vs 1.1%) [17]. Major bleeding rates with fondaparinux, 7.5 mg daily, are similar to those of enoxaparin LMWH (1.1% vs 1.2%) when these agents are used for the treatment of DVT [18].

Although prophylactic doses of heparins are lower than treatment doses, bleeding rates might be higher because the drugs are generally given to patients having major surgery. A meta-analysis that pooled data from 27 trial groups encompassing 1745 patients undergoing total hip replacement and receiving prophylactic doses of UH calculated that 2.6% had clinically important bleeding [19]. Double-blind, randomized trials that administered UH, 5000 IU every 8 hours, reported major bleeding in 2.5% of patients undergoing general surgery, 1.5% having colorectal operations, and 2.9% having elective cancer surgery [20−22]. However, higher bleeding rates with UH prophylaxis have been reported in other trials of patients undergoing major surgery (4.8%, 6.1%) [23,24].

A trial comparing prophylactic LMWH with UH after elective hip replacement reported similar rates of major bleeding: 4% in patients given enoxaparin,

30 mg every 12 hours, and 6% in those given UH, 5000 U every 8 hours [25]. Other trials in total hip arthroplasty recorded major bleeding in 2% of patients receiving dalteparin and 2.8% of those given tinzaparin [26,27]. A comprehensive examination of studies comparing LMWH with UH found no convincing evidence of a difference in bleeding between the two products, and similar rates of bleeding complications among the individual LMWHs [28]. An analysis of trials using fondaparinux, 2.5 mg daily, for thromboprophylaxis after major orthopedic surgery found a major bleeding rate of 2.7%, not significantly different from that of enoxaparin, 1.7% [29] (Table 5.2).

If treatment doses of LMWH are given during pregnancy, the risks of antepartum and postpartum hemorrhages are 1.5% and 2%, respectively [30]. Other individuals at increased risk of bleeding are those with thrombocytopenia or renal failure. Full doses of LMWHs are usually tolerated by patients with platelet counts $\geq 50,000/\mu L$, but giving only half the dose is suggested if counts are between $25,000/\mu L$ and $50,000/\mu L$, and discontinuing anticoagulants or transfusing platelets if the counts are $<25,000/\mu L$ [31]. Dose adjustments are imperative for patients with renal failure. An analysis of 10,687 hospitalized patients given enoxaparin for non-ST-segment elevation acute coronary syndromes reported that the 19% receiving doses >10 mg higher than recommended had an increased risk for death (5.6% vs 2.4%) [32]. A meta-analysis of 12 studies and 4971 patients noted significantly higher bleeding rates in those with creatinine clearances ≤ 30 mL/minute as compared to those with normal renal function (5.0% vs 2.4%, P = 0.013); adjusting the LMWH dose in the patients with renal impairment eliminated the differences in bleeding rates [33]. The incidence of major bleeding when fondaparinux is used for surgical prophylaxis is 1.9%, but in patients weighing <50 kg, it is 5.4%, and in those with kidney impairment (glomerular filtration rate 30—50 mL/minute), it is 3.8%.

Heparin-associated bleeding most commonly takes the form of hematomas at subcutaneous injection sites (Fig. 5.1), but wound hematomas occasionally occur in patients receiving anticoagulant prophylaxis for major surgery. Other sites of bleeding are the gastrointestinal, genitourinary, and respiratory tracts or into the retroperitoneal, pericardial, and intracerebral spaces. Entrapment of the femoral nerve by an expanding psoas muscle or retroperitoneal hematoma is recognized by weakness of the quadriceps, diminished patellar reflexes, and sensory loss over the anteromedial aspect of the leg [34]. Other sites of bleeding are muscles, joints, and even the eye [35].

In 1965, bilateral adrenal hemorrhage developing during the first 10 days of heparin therapy was described in 10 patients [36]. Their symptoms were diffuse abdominal or back pain, listlessness, and weakness, and the laboratory findings were an elevated serum potassium and reduced serum sodium and glucose. At autopsy, the medulla and inner cortex of the adrenals were found to be completely replaced by hematomas and the outer cortex was necrotic. Should this diagnosis be suspected today, it would be confirmed by computerized tomography and magnetic resonance imaging. In most patients,

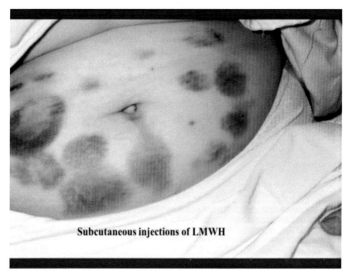

Subcutaneous injections of LMWH

FIGURE 5.1   Hematomas due to daily injections of LMWH.

it is likely that the adrenal involvement is due to heparin-induced thrombocy-topenia rather than anticoagulant-induced bleeding.

Spinal hematoma is a rare but serious complication of spinal/epidural anes-thesia in patients receiving anticoagulants, and has occurred in patients receiv-ing UH as well as LMWH. The estimated frequencies with UH are <1 in 5000, and with enoxaparin, 1 in 3100 with epidural anesthetics and 1 in 41,000 with spinal anesthetics [37]. It is more common in elderly patients with renal insufficiency, and anesthetic factors that increase the risk are traumatic needle/catheter placement, epidural as opposed to spinal technique, and having an indwelling epidural catheter during LMWH administration [38]. Spinal and epidural hematomas can compress the spinal cord, causing paraplegia. Because of the seriousness of this complication, the U.S. Food and Drug Administration (FDA) has required that every package insert for heparins carry a warning about the risk of spinal/epidural hematoma with neuraxial anesthesia (Fig. 5.2).

The American Society of Regional Anesthesia (ASRA) evidence-based guideline for UH recommends that needle/catheter placement be delayed for 4−6 hours after the last dose of continuous intravenous UH, and the aPTT be in the normal range [39]. UH is resumed one hour after the procedure. For patients receiving prophylaxis with subcutaneous UH in daily doses of 5,000 U every 8−12 hours, catheter placement is delayed for 4−6 hours after the last dose; when higher doses of subcutaneous UH have been adminis-tered, catheter placement is delayed for 24 hours. Neuraxial anesthesia may be continued in patients receiving low dose UH, but it should be used with utmost caution and aggressive neurologic monitoring in patients receiving higher doses; catheter removal is safe 4−6 hours after the last dose of UH.

> WARNING: SPINAL/EPIDURAL HEMATOMA
> Epidural or spinal hematomas may occur in patients who are anticoagulated with low molecular weight heparins (LMWH) or heparinoids and are receiving neuraxial anesthesia or undergoing spinal puncture. These hematomas may result in long-term or permanent paralysis. Consider these risks when scheduling patients for spinal procedures. Factors that can increase the risk of developing epidural or spinal hematomas in these patients include:
> *Use of indwelling epidural catheters
> *Concomitant use of other drugs that affect hemostasis, such as non-steroidal anti-inflammatory drugs (NSAIDS), platelet inhibitors, other anticoagulants
> *A history of traumatic or repeated epidural or spinal punctures
> *A history of spinal deformity or spinal surgery
> *Optimal timing between the administration of the heparin and neuraxial procedures is not known
> Monitor patients frequently for signs and symptoms of neurological impairment. If neurological compromise is noted, urgent treatment is necessary.
> Consider the benefits and risks before neuraxial intervention in patients anticoagulated or to be anticoagulated for thromboprophylaxis.

FIGURE 5.2   Boxed warning about spinal/epidural hematoma.

The ASRA Guidelines for LMWH recommend that catheter insertion should occur no earlier than 12 hours after a prophylactic dose, and 24 hours after the last treatment dose. Postoperative LMWH is initiated when adequate hemostasis has been secured and no sooner than the day after surgery. LMWH is given at least 12 hours after catheter placement, and held for 4 hours prior to catheter removal. However, when LMWH is given twice daily, anti-Xa levels >0.2 U/mL have been recorded at the time of planned catheter removal; [40] therefore, dosing should be decreased to once daily prior to removal. Indwelling neuraxial catheters may remain in place if LMWH is given only once daily, but doses should be held for 12 hours prior to catheter removal and resumed 4 hours later; other medications that alter hemostasis should be avoided. For example, ketorolac (Toradol®) is commonly used for analgesia in post-operative patients; it is a non-steroidal anti-inflammatory drug (NSAID) that relieves pain but inhibits platelet function. In a rabbit model of hemostasis, we observed that blood loss was significantly increased in animals receiving both LMWH, 100 U/kg, and ketorolac, 1 mg/kg, but not with smaller doses of the NSAID [41]. In human volunteers, platelet aggregation was inhibited by ketorolac and there was a significant interaction between ketorolac and LMWH in prolonging the skin bleeding time (with LMWH alone, 6–7 minutes; ketorolac, 10.5 minutes; and both drugs, 14 minutes) [42]. When antiplatelet medications are combined with LMWH, the risk for spinal hematoma is increased [39].

Fondaparinux presents more of a problem for neuraxial catheter placement and removal because of its long half-life (17 hours). In a clinical trial, catheters were removed 36 hours after the last fondaparinux dose and the anticoagulant was resumed 12 hours after catheter removal; there were no epidural or spinal hematomas [43]. The ASRA Guidelines recommend that indwelling neuraxial catheters be avoided in patients receiving prophylaxis with fondaparinux [39].

Management of bleeding

Management must balance aggressiveness in controlling bleeding with the enhanced risks of thrombosis. Life-threatening hemorrhages require immediate cessation of the anticoagulant, administration of a reversal agent, and transfusion of blood products. Clinically relevant non-major bleeding, defined as bleeding requiring medical intervention, might be managed by temporarily pausing anticoagulant therapy until the bleeding site has been controlled. If anticoagulants must be withheld for more than a few days in patients at high risk of recurrent venous thromboembolism, a retrievable vena cava filter can be inserted and removed after anticoagulants are resumed [19]. Agents used for the control of anticoagulant-associated bleeding are discussed below; a recent review of this topic includes sections describing supportive care and the comprehensive management of intracranial and gastrointestinal hemorrhages [44].

The anticoagulant activity of UH is almost completely reversed by protamine sulfate, a positively-charged arginine-rich peptide that binds to the negatively-charged heparin [45]. The dose is dependent on the timing of the last heparin dose, calculated in increments of 100 U; it is 1 mg if within 30 minutes of heparin dosing, declining to 0.25 mg if >120 minutes since the last dose [46]. Protamine (10 mg/mL) is infused slowly intravenously at a rate ≤ 5 mg/minute; the total dose should not exceed 50 mg. The complications of protamine therapy include hypotension as well as hypersensitivity reactions that are 10-fold more frequent in people with fish allergies or previous exposure to protamine or protamine-containing insulin products than in those without such exposure (0.6% vs 0.06%) [47]. The aPTT can be monitored following protamine administration, and if it remains prolonged or bleeding persists, dosing can be repeated.

Although the antithrombin activity of LMWHs is neutralized by protamine, the anti-Xa activity of LMWHs is incompletely (65%) reversed [48]. If given with 3−4 hours of LMWH, the protamine dose is 1 mg per 1 mg enoxaparin or 100 U of dalteparin, nadroparin, or tinzaparin. If serious bleeding continues, a second dose of 0.5 mg can be infused. The anti-Xa activity should be monitored because clinical responsiveness to protamine has been variable [49]. Bleeding not responsive to protamine might be controlled by recombinant activated factor VII (rVIIa), a potent procoagulant that has been reported to control bleeding in a patient receiving LMWH [50]. An initial dose that is half or less than that used for the treatment of factor VIII inhibitors (40 µg/kg is suggested) might reduce the risk of thromboembolic complications but still be effective in establishing hemostasis.

Andexanet, an agent that binds to the active site of factor Xa inhibitors, was approved recently for the management of patients with life-threatening or uncontrolled bleeding due to rivaroxaban or apixaban. The agent has also been evaluated for the management of bleeding associated with LMWHs. In healthy volunteers given enoxaparin, 40 mg daily for 6 days, anti-Xa levels were reduced from 0.36 IU/mL to 0.12 IU/mL, and inhibition of thrombin generation was reversed [51]. A clinical trial that included 16 patients with major bleeding attributed to enoxaparin found that andexanet, 800 mg bolus followed by an infusion of 8 mg/minute for 2 hours, decreased anti-Xa activity from 0.48 to 0.15 ng/mL and provided excellent or good hemostasis in 14 (87%) [52]. Andexanet is not FDA-approved for bleeding due to LMWHs, and the package insert has a black box warning of thromboembolic and ischemic risks, cardiac arrest, and sudden death. Further evaluation of this or similar agents is needed.

The anticoagulant activity of fondaparinux persists for many hours because of its long half-life, and it is not reversed by protamine. Because there is no specific reversal agent, clotting factor concentrates are usually administered to control major bleeding. A single intravenous bolus of rVIIa, 90 µg/kg, given 2 hours after healthy men received a 10 mg dose of fondaparinux, returned the aPTT to the normal range for up to 6 hours [53]. An observational study reported rVIIa controlled major bleeding attributed to fondaparinux and platelet inhibitors in four of 8 patients; the individuals resistant to rVIIa had baseline anti-Xa activity >1.0 U/L [54]. Another procoagulant product, activated prothrombin complex concentrate (aPCC), was shown to improve the endogenous thrombin potential in 6 volunteers given fondaparinux, but no improvement was observed with rVIIa [55]. Based on this study, the Guidelines from the Neurocritical Care Society and the Society of Critical Care Medicine recommend that patients with intracranial hemorrhage initially receive an aPCC, 20 IU/kg, to reverse the anticoagulant activity of fondaparinux, and suggest that rVIIa be given if an aPCC is contraindicated or not available [56].

Patients treated for anticoagulant-associated bleeding remain at risk for thrombosis, so resumption of antithrombotic therapy should be considered as soon as clinical circumstances permit. The gastrointestinal tract is the most common site for major bleeding; re-bleeding is more likely if anticoagulants are resumed <2 weeks rather than >2 weeks after a hemorrhage (hazard ratio, 2.4) [57]. Resuming anticoagulants at the time of hospital discharge appears to reduce the risk of thromboembolic events without increasing the incidence of recurrent gastrointestinal hemorrhage [58]. Because there are few studies of re-bleeding in patients with heparin-associated intracranial hemorrhage (ICH), data are extrapolated from experience with warfarin. These reports suggest that resumption of anticoagulation within the first month after discharge for an ICH is associated with few thromboembolic events and a low risk of recurrent bleeding [59].

Heparin-induced thrombocytopenia

In 1969, a case report described a patient with prostate cancer and thrombocytopenia who was given heparin for treatment of an intravascular coagulation syndrome [60]. After completion of an initial course of UH, the platelet count had increased to 115,000/μL, but the fibrinogen declined, leading to a resumption of heparin therapy. The fibrinogen concentration promptly rebounded, but the platelet count fell to 10,000/μL and the heparin was discontinued. *In vitro* studies showed that the addition of heparin to the patient's platelet-rich plasma induced a 40% reduction in the platelet count, but had no such effect on the platelet-rich plasma of four controls. This was the first report to use the term, "Heparin-induced Thrombocytopenia," although the thrombotic manifestations of this syndrome had been recorded previously in surgical patients [61]. Subsequent publications by Rhodes, Dixon, and Silver [62,63] recognized that both the thrombocytopenia and thrombosis were features of an adverse reaction to heparin, and that cessation of the anticoagulant was "imperative for the successful management of afflicted patients."

Investigators observed that the addition of heparin to patient platelet-rich plasma induced platelet aggregation (Fig. 5.3) [64]. The detection of heparin-dependent antiplatelet antibodies in the plasma indicated an immune basis for the syndrome [62,65]. Study of serum from a patient with heparin-induced thrombocytopenia (HIT) suggested that the antibody was directed against a complex of heparin and a platelet component [66]. It was suspected that binding of this complex to platelets triggered the platelet release

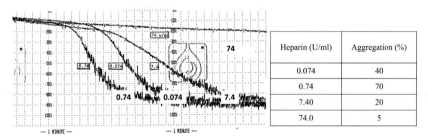

Heparin (U/ml)	Aggregation (%)
0.074	40
0.74	70
7.40	20
74.0	5

FIGURE 5.3   Heparin-induced platelet aggregation. Various concentrations of heparin, ranging from 0.074 to 74 U/mL, were mixed with platelet-rich plasma from a patient with heparin-induced thrombocytopenia, and the extent of platelet aggregation recorded. Note that a specific heparin concentration (0.74 U/mL) induced maximum aggregation; this dependency on the concentration of the antigen (heparin) is used to confirm the immunologic character of the reaction.

reaction, platelet aggregation, and thrombocytopenia, as well as contributed to thrombus formation. It was not until 1992 that platelet factor-4 (PF4 or CXCL4, a chemokine abundant in platelet alpha-granules) was identified as the platelet component of the tri-molecular complex [67]. Confirmation came from a study of platelets congenitally-deficient in PF4 (gray platelet syndrome); adding heparin and IgG from a patient with HIT to these platelets released serotonin only when purified PF4 was provided [68]. Other studies showed that HIT-antibodies react with heparin bound to endothelial cells or heparan sulfate elaborated by these cells, suggesting that endothelial injury might contribute to thrombus development [69]. In 1994, Warkentin et al [70] reported that HIT-antibodies stimulated the formation of platelet-derived procoagulant microparticles both *in vitro* and *in vivo*, providing a compelling explanation for the thrombotic complications associated with this disorder.

The incidence of HIT compiled from the early literature was 3.4%, with a range of 1.3%−11.6% [71−73]. This broad range is probably related to the variability in heparin preparations and the diverse patient populations evaluated. In the 1980s, bovine heparin was in common use and in some studies appeared more likely than porcine heparin to induce thrombocytopenia [74−76]. Clarification came with the observation that a specific lot of beef lung heparin induced platelet antibody formation and thromboembolic complications in patients undergoing coronary artery bypass surgery [77]. Thus, it seems likely that the heterogeneity of heparin products manufactured from bovine lungs accounted for much of the reported difference in HIT incidence. Virtually all heparin fractions in current use are extracted from porcine intestinal mucosa.

The risk factors for HIT are shown in Table 5.3. The duration of heparin exposure is a key factor; 5−7 days are usually required for a drug to elicit antibodies unless prior exposure has occurred. The longer the exposure, the

TABLE 5.3 Risk factors for heparin-induced thrombocytopenia.

Risk factor	Description
Dose of heparin	Treatment > prophylactic
Initial exposure	> 5 days
Re-exposure	< 100 days
Product	UH > LWMH > fondaparinux
Platelet FcγRIIA receptor	R isoform of H131R receptor
Indication	Cardiac & orthopedic surgery major trauma
Co-morbidities	Autoimmune disease
Gender	Female > male

greater risk of HIT; for example, the relative risk for exposures of 5–10 days is 5.51, but for >10 days, it is 7.66 [78]. During recovery from HIT, heparin-dependent antibodies fall to undetectable levels after 50–85 days and are unlikely to reappear after re-exposure to heparin if >100 days have elapsed [79,80]. A study showed that 17 patients examined 4.4 years after an episode of HIT had no demonstrable platelet-activating HIT-antibodies [81]. They were subsequently given heparin intraoperatively for cardiac or vascular surgery. Seroconversion occurred in 65% but only one patient developed HIT, and not until seven days postoperatively. Another report suggested that HIT-patients with recovered platelet counts but persistently positive PF4/heparin antibody assays can tolerate brief re-exposure to UH during heart transplant surgery if they have negative sensitive functional assays for HIT antibodies using washed platelets [82].

Several studies have shown that the risk of HIT is considerably greater with UH than with LMWH, and rare with fondaparinux.; the incidence with prophylactic doses of UH is 2.6%, with LMWH 0.2%, and it is extremely rare with fondaparinux [83–86]. Patients receiving prophylaxis for cardiac and orthopedic surgery are at relatively high risk. A study of 754 patients reported that HIT-antibodies developed in 20% of cardiac and 9.3% of orthopedic patients given UH, but the frequency of HIT was higher in orthopedic than cardiac patients (4.9% vs 1%) [87]. HIT-antibodies were detected in 3.2% of orthopedic patients receiving LMWH but only 0.9% experienced HIT. A meta-analysis encompassing >7000 patients confirmed that the risk of HIT with LMWH was only one-tenth of that with UH [84]. Although antibodies can be detected in orthopedic surgery patients receiving fondaparinux, they fail to react *in vitro* with complexes of fondaparinux and platelet factor 4, and none of the 1377 patients exposed to this synthetic heparin developed HIT [88]. The non-reactivity of fondaparinux was confirmed by a prospective, blinded study of 39 HIT sera; only 3.3% of 91 assays gave positive reactions [89].

The risk of HIT is also influenced by the platelet Fcγ receptor [90]. In a mouse model, thrombocytopenia and thrombosis resulted when this receptor was activated by PF4/heparin-IgG immune complexes [91]. The frequency of the FcγRIIa-131RR allotype is significantly increased among patients with HIT and thrombosis (P = 0.036) [92]. On the other hand, a lower risk of HIT is associated with the 276P and 326Q alleles of CD148, a protein that regulates FcγRIIA signaling [93].

Individuals are more likely to develop antibodies if tissue destruction produces damage-association molecular patterns ("danger signals") that trigger activation of antigen-presenting cells [94]. This phenomenon could account for HIT developing in patients undergoing major cardiac or orthopedic surgery exposed to only prophylactic doses of heparin, and why HIT is confined to those with severe, but not minor trauma [95]. The incidence of HIT is lower in medical patients receiving UH prophylaxis (0.2%–0.8%), but is higher in those with autoimmune disease, gout, and heart failure [78,96]. Patients with HIT are more likely to have other diseases of immune dysregulation such as the antiphospholipid antibody syndrome, systemic lupus erythematosus, and autoimmune thyroiditis, suggesting a common pathogenesis [97]. The antibodies detected in some of these patients might be PF4 autoantibodies rather than heparin-induced antibodies, accounting for false-positive HIT tests using PF4 as target antigen [98]. Patients on chronic intermittent dialysis are regularly exposed to heparin and as many as 10.3% have HIT antibodies, but thrombocytopenia is uncommon [99,100]. On the other hand, many patients in intensive care units have thrombocytopenia and are receiving heparin, but the frequency of HIT is only 0.3%−0.5% [101]. Heparin exposure, heparin antibodies, and thrombocytopenia might co-exist in a patient but are not sufficient for a diagnosis of HIT in the absence of specific clinical features. Lastly, female gender is a risk factor for HIT (odds ratio, 2.37), and the risk is greater with UH than with LMWH [102].

Pathogenesis

Heparin infused intravenously in doses >50,000 IU promotes transient thrombocytopenia without clinical sequelae, and heparin added to platelet-rich plasma *in vitro* induces platelet aggregation and enhances aggregation induced by other agents [103]. In contrast, HIT occurs when heparin binds to PF4 on the platelet membrane. Antibodies target macromolecular complexes of PF4 and heparin; binding of these complexes to platelet receptors triggers platelet activation [104−108]. Furthermore, HIT antibodies bind PF4/heparin macromolecules on monocytes, neutrophils, and endothelial cells, activating these cells as well [109]. Activated platelets aggregate and are incorporated into "white" clots, contributing to the thrombocytopenia and thrombosis characteristic of HIT [110].

HIT-antibodies develop with subcutaneous as well as intravenous heparin exposure, and are able to activate platelets and trigger thrombosis [111,112]. *In vitro*, the addition of HIT-IgG antibodies and heparin can release serotonin from normal ^{14}C-serotonin-labeled platelet-rich plasma; [113,114] this reaction requires platelet FcγRII receptors [94]. The HIT-antibodies induce clustering, oligomerization, and cross-linking of PF4 tetramers on the platelet surface, exposing binding sites for additional antibody molecules [115,116]. The binding strength to platelet receptors is greatest with PF4 tetramers and UH, intermediate with LMWH, and weakest with fondaparinux; the latter forms only single PF4/heparin molecules [117]. HIT is self-propagating; a dynamic cycle is initiated because platelets activated by binding HIT-antibodies release additional PF4, form new multimolecular complexes that attach to the surface of platelets and other cells, and perpetuate the syndrome [118].

Marginal zone B-cells and thymic-derived T-cells are required for the generation of antibodies to the PF4/heparin complexes [119−121]. Although anti−PF4/heparin antibodies often appear in individuals exposed to heparin, they are present in low titer; B-cell tolerance to PF4/glycosaminoglycan (GAG) complexes probably accounts for the relative rarity of HIT [122]. In patients that develop HIT, IgG antibody levels are higher, become detectable 4 days after exposure to heparin, and precede the decline in platelet count by 2 days; levels of IgM or IgA antibodies are similar to those in HIT-negative controls [123]. Antibody concentrations reach a peak in 10 to 12 days and then decline despite continued heparin exposure, suggesting that B-cell activation is limited by developing tolerance [124]. Antibody assays generally become negative 5−6 weeks after withdrawal of heparin [125].

Thrombi in patients developing HIT are usually located at sites of vascular injury-arteries in patients undergoing cardiac surgery and the deep veins in those having orthopedic procedures. Several mechanisms account for the high frequency of thromboses. The HIT antibodies bind to equimolar complexes of PF4 and heparin on the platelet surface, more so if the platelets are activated [126,127]. The binding site is the FcγRIIa receptor; individuals homozygous for the 131RR receptor genotype have enhanced HIT-antibody binding and are at increased risk of thrombosis [128]. Cross-linking of the receptors mobilizes calcium, mediated by cytoplasmic 12(S)-lipoxygenase; [129] synthesis of thromboxane A_2 induces platelet aggregation and the release of microparticles, contributing to thrombus formation [130,131]. Thrombin translocates platelet phosphatidylserine from the inner to the outer membrane, facilitating the retention of procoagulant proteins on the surface and greater platelet thrombogenicity [132]. This can be demonstrated by noting enhanced thrombin generation when UH and HIT plasma are added to normal platelet-rich plasma [133].

HIT antibodies can bind to PF4 bound to linear polyanions other than heparin if the negative charges on these polyanions are spaced about 0.5 nm apart along the molecular backbone and are of sufficient length to span about 40% of the circumference of the PF4 tetramer [134]. Chondroitin sulfate (CS) is the most abundant glycosaminoglycan (GAG) on platelets and is present on other cells as well. A subset of HIT antibodies recognizes PF4-CS complexes on the surface of platelets and PF4-GAGs on monocytes [109,135]. These antibodies form multimolecular complexes on monocytes that bind to FcγR1 and RIIA receptors, releasing microparticles expressing tissue factor and thrombin [136−138]. HIT antibodies also bind to complexes of PF4-GAGs on endothelium, activating these cells and releasing tissue factor [139]. Interchange of PF4-GAG complexes between platelets, monocytes, and endothelial cells modifies the severity of the thrombocytopenia and the propensity to thrombosis [140]. The ability of pathogenic HIT-antibodies to bind to PF4-GAG complexes and activate cells explains the continuing risk of thrombosis even after heparin has been discontinued [141]. It also could account for the observation that HIT-antibodies are occasionally detected in patients, such as those with acute coronary syndromes, in the absence of exposure to heparin; [142] these are most likely induced by PF4-GAG complexes. Patients with acute coronary syndromes exhibit strong inflammatory responses accompanied by T-cell dysregulation and B-cell activation, with enhanced antibody formation [143].

Protein C, a potent natural anticoagulant, is activated by a complex of thrombin and thrombomodulin (TM). CS is a component of glycanated TM; PF4 released from platelets and other cells can bind to CS on TM [144]. In consequence, TM becomes a target for HIT antibodies, and the generation of activated protein C is inhibited, contributing to the thrombotic milieu [145]. In addition, plasma from patients with HIT induces the formation of leukocyte-platelet aggregates and activates neutrophils [146,147]. HIT immune complexes bind to neutrophil FcγIIA receptors and induce formation of neutrophil extracellular traps (NETs) [148]. The extruded NETs with their DNA and citrullinated histone H3 are incorporated into thrombi; in experimental models, neutrophil depletion abolishes thrombus formation, demonstrating the key role of these cells in HIT pathogenesis [149].

Interleukin-8 is a cytokine produced by phagocytes and mesenchymal cells exposed to inflammatory stimuli; it is a potent activator of neutrophils [150]. HIT-antibodies and PF4-heparin complexes induce monocytes and immune cells bearing toll-like receptors to synthesize and secrete IL-8, possibly representing a misdirected host defense response [151−153]. Evidence to support this view comes from studies demonstrating that PF4 binds to a variety of bacteria, eliciting antibodies that recognize PF4/polyanion complexes, including PF4/heparin [154]. These antibody complexes are found in people with periodontal disease, independently of other markers of inflammation [155]. It is conceivable that the PF4/heparin complexes that form when heparin is given to individuals with chronic infections might be misinterpreted as evidence of a bacterial invasion, producing an anamnestic response with elaboration of HIT-like antibodies. Such a scenario would be consistent with the appearance, as early as 4−5 days after exposure to heparin, of IgG rather than IgM antibodies, followed by a relatively rapid decline in antibody titers as tolerance returns [156].

PF4 is a ubiquitous actor in a variety of inflammatory disorders; for example, in patients with scleroderma, it assembles DNA into complexes to amplify interferon-α production by dendritic cells [157]. These observations raise the question of why HIT doesn't occur with every exposure to heparin. Rauova et al. [158] explain that PF4 rapidly dissociates from cell-associated GAGs, and heparin removes all PF4 from the cell surface leaving no target for antibodies. On the other hand, patients with the risk factors shown in Table 5.3 usually have ongoing platelet activation and continuing PF4 release; the formation of PF4/CS complexes on the cell surface provides a ready target for antibodies and eventuates in the thrombocytopenia and thrombosis characteristic of HIT.

Diagnosis

The defining feature of new onset HIT is thrombocytopenia. The timing and extent of the platelet decline are well-characterized in patients exposed to heparin during cardiopulmonary bypass: the decline in platelet count begins 5−10 days after heparin exposure and the extent of the decline is >50% [159]. While the nadir is usually ≥ 20,000/μL, occasionally the platelet count remains above 100,000/μL. In contrast, persistent thrombocytopenia beginning before postoperative day 5 is rarely due to HIT [160]. If a previous heparin exposure has occurred within 30 days, the platelet count might fall in a day or less. A few intriguing variations in HIT presentations have been reported-delayed onset, counts remaining within the normal range, and HIT without obvious exposure to heparin-and are discussed in the next paragraph.

Delayed onset HIT can occur up to 40 days after exposure to heparin, and patients might have already been discharged from the hospital [161]. These individuals have the usual features of HIT and have been shown to have higher titers of HIT-antibodies than patients with typical HIT [162]. There are a few reports of well-defined HIT without thrombocytopenia; in some patients, elevated platelet counts declined by 50% but were still within the normal range [163], while in others, the low normal platelet counts increased by 50% when heparin was replaced by another anticoagulant [164]. Lastly, patients have been reported with thrombocytopenia and thrombosis typical of HIT, but without heparin exposure [165]. They often have diseases that might release PF4, and binding of the PF4 to other negatively-charged molecules, such as GAGs or nucleic acids, could elicit antibody formation and produce the clinical features of HIT [166−168]. Patients without proximate heparin exposure but with thrombocytopenia, thrombosis, and platelet-activating antibodies are said to have spontaneous HIT [167].

The other major characteristic of HIT is the appearance of new venous thromboemboli, myocardial infarction, or ischemic stroke [169]. In addition, central venous catheters might become occluded [170] and carotid endarterectomy sites thrombosed; [171] thrombi often develop at sites of localized vascular injury. A rare presentation is bilateral adrenal hemorrhage secondary to thrombosis of the adrenal veins [172]. Another serious complication of HIT is venous limb gangrene, usually in association with warfarin therapy [173]. A common scenario is the appearance of distal limb cyanosis and necrotic skin bulli in a patient with HIT in whom heparin has been discontinued but warfarin has either been initiated or continued. The distal veins are thrombosed, the International Normalized Ratio (INR) is often above the therapeutic range, and protein C activity is severely depressed. The use of vitamin K-antagonists should be avoided in patients with acute HIT, and vitamin K administered to raise protein C levels if patients have received a vitamin K antagonist [174].

Necrotic skin lesions have been reported in patients with HIT, even in the absence of vitamin K antagonists [175,176]. This complication occurs after exposure to LMWH as well as UH [177], but can be difficult to distinguish from other manifestations of heparin allergy. For example, an observational study described 87 consecutive patients with heparin-induced skin lesions; these were attributed to delayed type IV hypersensitivity reactions and only one of the patients met diagnostic criteria for HIT [178,179]. Other clinicians have described Arthus-like reactions (local vasculitis and deposition of immune complexes) occasionally complicated by skin necrosis; some of these patients might have had HIT [180−182].

The *4Ts score* awards points for Thrombocytopenia, Timing of platelet count fall, Thrombosis or other sequelae, and oTher causes of thrombocytopenia (Table 5.4). The negative predictive value of a low score (1−3 points) was 0.99, suggesting the absence of HIT and need to perform additional testing; conversely, patients with ≥4 points underwent further evaluation for HIT [183]. A prospective evaluation of 526 patients suspected of having HIT combined the 4Ts score with a rapid agglutination test for PF4/heparin antibodies, and noted that the sensitivity and specificity of the 4Ts score were only 81.3% and 63.8%, respectively; however, HIT could be confidently excluded in all patients with low/intermediate 4Ts scores and a negative PF4/heparin agglutination test [184].

Another clinical diagnostic system is the HIT Expert Probability (HEP) score [185]. It is more complex, with eight items scored. A prospective comparison with the 4Ts score in 310 patients with suspected HIT (prevalence 14.7%) noted that both tests were quite sensitive but had low specificity (HEP, 35.7%; 4Ts, 32.9%) [186]. They had similar areas under the receiver operating curve (AUC), but the HEP score had a higher AUC than the 4Ts score in patients in the intensive care unit (0.86 vs 0.79, P = 0.03), and might be preferable for that population.

TABLE 5.4 4Ts scoring system for pretest probability of HIT (low: 1−3 points; intermediate: 4−5 points; high: 6−8 points) [184].

Thrombocytopenia (1 point if count has fallen 30%−50% or nadir is 10−19, and 2 points if fall is >50% and nadir ≥20);

Timing of platelet count fall (1 point if onset suspected to be ≥5 d or ≤1 d if prior exposure to heparin within past 30−100 d, 2 points if definite onset 5−10 d or ≤1 d with prior heparin within 30 d);

Thrombosis or other sequelae (1 point for progressive or recurrent thrombosis, non-necrotizing skin lesions, or suspected thrombosis, 2 points if new thrombosis, skin necrosis, acute reaction to heparin bolus);

oTher causes of thrombocytopenia (1 point if possible, 2 points if none apparent).

Laboratory testing

Guidelines suggest that patients receiving UH (medical or obstetrical patients) or LMWH (for major surgery or trauma) have platelet counts every other day beginning the day heparin is initiated and continuing until day 14 [187]. Laboratory testing is not recommended for those with a low 4Ts score (≤3 points), and heparin might be continued. On the other hand, higher 4Ts scores are followed-up with platelet activation studies and/or immunoassays.

As early as 1975, Fratantoni et al. [188] reported that serum from a patient who became thrombocytopenic while receiving heparin induced the release of H^3-serotonin from normal platelets. The ^{14}C-serotonin-release assay (SRA) for the diagnosis of HIT was described in 1986 and included two concentrations of heparin: 0.1 U/mL (a therapeutic level) and 100 U/mL; the latter was necessary to exclude non-specific release [189]. A positive test is defined as ≥20% serotonin release with 0.1 U/mL heparin and <20% release with 100 U/mL using heat-inactivated patient citrated plasma or serum; however, nearly all clinically significant antibodies produce >50% serotonin release with the lower heparin concentration [81]. The SRA is >90% specific for HIT and is considered the gold-standard test for this disorder, but at present it is performed by only a few laboratories.

Another test for the detection of pathogenic HIT antibodies is the PF4-dependent P-selectin expression assay (PEA) [190]. The method is based on flow-cytometry; normal platelets pretreated with 3 concentrations of PF4 are incubated with suspected HIT plasma and the surface expression of P-selectin, an indicator of platelet activation, is measured. An evaluation of 91 patient samples found that the test was more sensitive and had greater accuracy than the SRA.

The most commonly used laboratory test for HIT is an enzyme-linked immunosorbent assay (ELISA) similar to the one originally described by Amiral et al. [191] in 1992. In principle, the PF4-heparin complex is affixed to microtiter wells and patient serum added; HIT antibodies will bind to the PF4-heparin on the solid support. The washed wells are then incubated with anti-human globulin labeled with alkaline phosphatase, followed by p-nitrophenyl phosphate; the resulting optical density (OD) reading at 405 nm is proportional to the concentration of bound HIT antibody. Warkentin et al. [192] report that for every increase of 0.5 OD units, the risk of a strongly positive SRA increases by an odds ratio of 6.4; the probability of HIT antibodies reaches ≥50% only when the OD level is ≥1.40 units. Other workers observe that OD readings of 0.4, 1.3, and 2.0 correspond to low, intermediate, and high diagnostic thresholds for HIT [193]. Adding excess heparin to the sample decreases the OD by ≥50%, confirming the presence of HIT antibodies [194].

The particle gel immunoassay is a commercially-available technique that generates a result within 20−30 minutes; a negative test helps to exclude the diagnosis of HIT [195]. A more exacting method is the IgG-specific chemiluminescence assay; it is reported to be 98.8% sensitive, 98.5% specific, and in 98.4% agreement with the SRA. Test results are said to be available 30 minutes after sample preparation [196]. There is also an assay based on the detection of a platelet FcγRII receptor fragment that is produced only by HIT antibodies [197].

A negative electroimmunoassay test result is useful in ruling out HIT, but because immunologic assays are often sensitive to non-pathogenic HIT antibodies, a positive test is not always indicative of clinical HIT [198,199]. Positive ELISA results have been reported in 16.8% of patients with antiphospholipid antibodies (most had never been exposed to heparin) [200]. The clinical scoring systems should be combined with the laboratory assays to improve the pretest probability of HIT [183,186,201]. Also, the diagnosis of HIT after cardiopulmonary bypass is more secure if the interval of a biphasic platelet count from the time of bypass to the first day of suspected HIT is ≥ 5 days, and the bypass duration ≤ 118 minutes [202].

Treatment

Case Study: A 39 year old woman became pregnant after receiving fertility drugs, but developed a left subclavian vein thrombosis. She was given enoxaparin, 1 mg/kg twice daily, and did well until the 16th week of gestation when she was admitted with an ischemic stroke and was found to have a platelet count of 5000/μL. Imaging studies revealed thrombosis of the superior sagittal venous dural sinus. Heparin-induced thrombocytopenia with thrombosis (HITT) was suspected, the enoxaparin was discontinued, and danaparoid was administered. The HITT resolved and she eventually delivered a premature but healthy baby.

When HIT is suspected, all exposures to heparin should be discontinued immediately; these include line flushes with heparin [203], insertion of heparin-coated catheters, and use of heparin-containing dialysate. Because of ongoing risks of thrombosis, rapid-acting non-heparin anticoagulants should be started and are described in the following paragraphs. Laboratory tests are ordered to confirm the diagnosis, and the platelet count closely monitored. Patients are repeatedly evaluated for signs and symptoms of new thromboembolic events. Thromboses should not be treated with LMWH or coumarin derivatives, and platelet transfusions should be avoided if possible. The insertion of an inferior vena cava filter in patients with HIT is often complicated by vena cava thrombosis and is not recommended [187].

Argatroban, a direct thrombin inhibitor, is FDA-approved for the management of HIT. The drug reversibly binds to the active site of thrombin and prolongs the aPTT. A report in 1992 described a patient who developed a declining platelet count and clots in the dialyser when UH, and subsequently LMWH, was used for hemodialysis; substituting argatroban led to an increase in the platelet count and cessation of clot formation [204]. Approval of the anticoagulant was partly based on a clinical trial of 304 patients with HIT whose outcomes were compared to 193 historical controls [205]. A composite endpoint of new thromboses, amputation, or death occurred in fewer argatroban-treated patients than controls (25.6% vs 38.8%, P = 0.014); in patients with heparin-induced thrombocytopenia with thrombosis (HITT), the composite incidence was 43.8% vs 56.5% (P = 0.13). Thrombocytopenia resolved in 56% within 3 days and in 75% during the treatment course. Major bleeding occurred in 7%; other adverse effects were diarrhea (11%) and pain (9%).

The approved initial dose of argatroban is based on body weight and is 2 µg/kg/minute given as a continuous intravenous infusion. The baseline aPTT should be prolonged 1.5−3-fold (but not exceed 100 s) in samples drawn 2 hours after initiating the infusion; values within the therapeutic range should be attainable in ≥ 83% of patients within 4−5 hours. Infusion rates should be adjusted (but not by more than 10 µg/kg/min) to achieve a steady state aPTT within the therapeutic range, and the aPTT re-checked after each adjustment. The initial dose is reduced to 0.5 µg/kg/minute in patients with moderate to severe hepatic impairment, and to 0.75 µg/kg/minute in pediatric patients. A response to argatroban is usually observed within 3−5 days, with platelet counts rising above 100,000/µL and signs of thrombosis abating. The major adverse effect of the drug is bleeding; other reactions such as dyspnea, hypotension, and fever are infrequent.

An alternative to argatroban is bivalirudin, a 20-aminoacid synthetic peptide that reversibly binds to the catalytic site and anion-binding exosite of thrombin. The drug has a short half-life of only 25 minutes if renal function is normal; the half-life is approximately doubled if the creatinine clearance is ≤ 30 mL/minute [206]. A retrospective study evaluated bivalirudin in 461 patients with possible HIT (confirmed in 32%). The initial dose was 0.05−0.1 mg/kg/hour to achieve a 1.5−2.5-fold prolongation of the baseline aPTT; a median of 12 hours was required to attain therapeutic levels. Bivalirudin was continued for a median of 9 days (range, 5−15 days); 58.3% of patients had platelet recovery to >150,000/µL by treatment day 4 and only 4.6% had new thrombi [207]. However, major bleeding occurred in 7.6% and fatal hemorrhage in 1.5%. An advantage of treatment with bivalirudin is that its effect on coagulation dissipates within a few hours of discontinuing the infusion.

Two non-heparin anticoagulants approved by the FDA for acute HIT but no longer available in the U.S. are danaparoid and lepirudin. Danaparoid is obtained from pig intestinal mucosa and consists of 80% heparan sulfate and smaller amounts of dermatan sulfate and chondroitin sulfate. It has a molecular weight of 6000 daltons, a half-life of 24 hours, and is eliminated by the kidneys [208]. In therapeutic concentrations, it decreases PF4 binding to platelets, displaces PF4-heparin complexes from platelets, and interferes with interactions of HIT antibodies with platelets [209,210] In 1983, Harenberg et al. [211] reported that continuous infusion of the heparinoid over a two-week period was effective in resolving HIT in a patient with femoral vein thrombosis. Small case series suggested it was safe and effective [212], and a retrospective review of 230 patients with suspected HIT observed responses in 92.8% [213]. Fatalities attributed to the drug occurred in 3%, and cross-reactivity of HIT antibodies with heparin was reported in 10%. Cross-reactivity might have been responsible for a fatal myocardial infarction and ischemic stroke in a patient following four days of danaparoid therapy for HIT [214].

Danaparoid is administered intravenously by bolus injection of 2250 U, followed by 400 U/hour for 4 hours, 300 U/hour for 4 hours, and 150−200 U/hour for at least 5 days with a target anti-Xa level of 0.5−0.8 U/mL. It has been used in children with HIT and has been given during pregnancy, and does not appear to cross the placenta or contaminate breast milk. Danaparoid can be used for hemodialysis or cardiopulmonary bypass, employing the specific dosing schedules shown in Table 16.1 of reference 208. Although the drug was approved for the treatment of HIT in 1996, it was withdrawn from the U.S. market in 2002 because of short supply.

Lepirudin is an analog of hirudin, the thrombin inhibitor produced by the medicinal leech. The derivatives currently in use are manufactured by recombinant DNA technology. Lepirudin binds tightly to the catalytic site and exosite of thrombin, inhibiting both free and clot-bound thrombin. The half-life is 1.1−2 hours, and 90% of the drug is cleared and degraded by the kidney [215]. A dose-finding study in 82 patients with acute HIT reported the rapid achievement of aPTT values within the target range of 1.5−2.5 times baseline, and an increase in platelet counts in 88.7% of patients with HIT [216]. As compared to historical controls, rates of death, limb amputations, and new thromboembolic complications were decreased, but bleeding events were more frequent [217]. These outcomes were confirmed by studies enrolling many more patients [218], and the investigators were able to identify the factors that contributed to bleeding as mean lepirudin dose greater than 0.07 mg/kg/hour, long duration of therapy, and moderate to severe renal impairment [219]. The manufacturer recommends a bolus dose of 0.4 mg/kg given slowly intravenously, followed by a continuous infusion of 0.15 mg/kg/hour continued for up to 10 days. The aPTT is drawn 4 hours after the initiation of treatment and dosing is adjusted to achieve the target range of 1.5−2.5 times baseline. When larger doses of lepirudin are used for patients undergoing cardiopulmonary bypass, the ecarin clotting time is preferred for monitoring [220].

Bleeding, the principle adverse effect of lepirudin, might be reduced by modifying the dosing scheme shown in the package insert. This alternative dosing omits the initial bolus and begins with a continuous infusion of

0.08 mg/kg/hour in patients with normal renal function, with doses for renal impairment reduced to 0.04 mg/kg/hour for creatinine clearances of 30−60 mL/minute and 0.01−0.02 mg/kg/hour for creatinine clearance of <30 mL/minute [221]. With this regimen, only 1 thrombotic event occurred in 46 patients, platelet counts increased, and the major bleeding rate was just 8.7%. Other adverse effects of the anticoagulant aside from bleeding are allergic reactions, including anaphylaxis, and development of lepirudin antibodies. These form complexes with the drug and prolong its anticoagulant activity by delaying its elimination. Accumulation of these complexes requires adjustment of the lepirudin dose to maintain the aPTT within the therapeutic range. Although lepirudin was approved by the FDA for the treatment of HIT, it was subsequently withdrawn from the U.S. market in 2012 because of supply problems.

When to perform monitoring, the desired therapeutic range, and conditions requiring dose adjustments are shown in Table 5.5 for these four non-heparin anticoagulants used in the treatment of acute HIT.

After treatment of the acute episode, continuation of anticoagulation is necessary because the risk of thrombosis persists for weeks to months; candidate drugs are warfarin, fondaparinux, and direct oral anticoagulants. Warfarin is introduced slowly at a dose of 5 mg or less daily and gradually increased to achieve an INR between 2 and 3 [222]. Therapy with warfarin should overlap the parenteral agent for at least 5 days, to ensure adequate depression of the vitamin K-dependent clotting factors. The advantages of warfarin are its effectiveness if the INR is within the therapeutic range, general familiarity with its use, and low cost. A disadvantage is that the anticoagulants used to treat acute HIT prolong the prothrombin time so that the initial INR values are not solely

TABLE 5.5 Non-heparin anticoagulants used to treat acute heparin-induced thrombocytopenia.

Agent	Monitoring	Therapeutic range	Dose adjustment
Argatroban	aPTT at 2 h	1.5−3.0 × baseline	Liver impairment
Bivalirudin	aPTT at 2 h	1.5−2.5 × baseline	Renal impairment
Danaparoid	Anti-Xa level at 4 h	0.5−0.8 U/mL	Renal impairment
Lepirudin	aPTT at 2 h or ecarin clotting time	1.5−2.5 × baseline	Renal impairment

*aPTT = activated partial thromboplastin time.

reflective of warfarin's effect on coagulation. The adjustments to warfarin dosing that are required vary with each of the alternative anticoagulants: argatroban, bivalirudin, danaparoid, and lepirudin [222]. Another concern is that starting warfarin before HIT has resolved and platelet counts normalized risks the development of warfarin-induced venous limb gangrene, as previously noted. Lastly, patients with HIT are often very sensitive to warfarin because of concomitant illnesses, poor nutrition, and polypharmacy, which increases the difficulty of maintaining INRs in the therapeutic range.

An alternative to warfarin is the synthetic pentasaccharide, fondaparinux. *In vitro*, HIT antibodies do not bind to PF4-pentasaccharide complexes, and plasma samples from 25 patients with HIT induced platelet aggregation in the presence of UH but not the pentasaccharide [223,224]. A study of 16 patients given fondaparinux for acute SRA-confirmed HIT reported no new thromboses and only one major hemorrhage [225]. A retrospective review of fondaparinux therapy in 133 HIT patients noted that outcomes (thromboses and major bleeding) were no worse than in patients given argatroban or danaparoid; in fact, there were fewer deaths in the fondaparinux group (22.6% vs 38.7%, P = 0.007) [226]. A registry study of 84 patients with suspected HIT receiving fondaparinux recorded no HIT-specific complications or hospital deaths, although these occurred in about 10% of patients given argatroban, lepirudin, or danaparoid; bleeding rates were similar with all agents [227]. Reports of ongoing thrombocytopenia and thrombosis in HIT patients receiving the pentasaccharide have been very infrequent [228].

Patients with a resolving HIT episode can be transitioned readily from one of the intravenous alternative anticoagulants to subcutaneous fondaparinux for outpatient use. A cost-effectiveness study of anticoagulants for the management of HIT in the U.S. concluded that fondaparinux was considerably less expensive than argatroban or bivalirudin ($151 vs $1250 or $1466, respectively), and had a higher probability of averting adverse events [229]. Although fondaparinux has not been studied in randomized controlled trials, the American Society of Hematology (ASH) 2018 Guidelines include the pentasaccharide as a reasonable option in clinically stable patients at average risk of bleeding [187].

Direct oral anticoagulants (DOAC) are another option for patients with resolving HIT. Neither rivaroxaban nor dabigatran interact with PF4 or HIT antibodies *in vitro* [230]. An early note described platelet recovery in 5 to 6 days in two patients with HIT given rivaroxaban in doses of 15 mg twice daily [231]. Subsequently, Warkentin et al [232] reviewed their own experience and data from the literature and reported resolution of HIT with only one new thrombosis and no major hemorrhages in 46 patients given rivaroxaban. Similar outcomes were described in smaller numbers of patients managed with apixaban or dabigatran. The ASH 2018 Guidelines include DOACs as well as fondaparinux in their recommendations for stable patients with HIT; [187] if rivaroxaban is selected, dosing for patients with thrombosis is 15 mg twice daily for 3 weeks followed by 20 mg daily; for those without thrombosis, 15 mg twice daily until there is full platelet recovery, then 20 mg daily [233].

Other infrequently employed treatments for HIT are intravenous immunoglobulin (IVIg) and plasma exchange. A 1989 report described the resolution of HIT after the administration of IVIg to a patient with severe HIT complicated by venous thromboembolism, gastrointestinal bleeding, and progressive thrombocytopenia [234]. The effectiveness of this approach is probably due to IVIg's ability to saturate FcγIIA receptors, preventing HIT antibodies from activating platelets. *In vitro* studies showed that IVIg inhibited the expression of P-selectin by platelets exposed to PF4 and HIT antibodies [235]. IVIg has been used in patients with severe HIT relatively refractory to other treatments, and to prevent recurrence of HIT in patients with persisting HIT antibodies [236,237].

Patients with acute HIT can be prepared for urgent surgery by employing therapeutic plasma exchange. In view of the effectiveness of IVIg in achieving platelet quiescence, it has been suggested that plasma, rather than the usual 5% albumin, should be the replacement fluid for the exchange because it is source of immunoglobulin [238]. In patients having repeated plasma exchange, the platelet activation assay becomes negative prior to the enzyme immunoassay for HIT, and when that happens, safe performance of cardiac surgery using heparin might be possible [239]. Other agents with possible activity in acute HIT are inhibitors of platelet function, such as aspirin [240] and the platelet glycoprotein IIb/IIIa antagonists, SR121566A and abciximab [241,242]. Experimentally, inhibiting the tyrosine kinase, Syk, blocks FcγRIIA-mediated platelet activation and prevents thrombocytopenia and thrombosis in a HIT mouse model [243]. Another approach uses a bacterial protease to cleave the hinge region of the HIT IgG heavy chain; this also prevents antibody-induced HIT in a mouse model [244].

Skin reactions

In 1990, Bircher et al. [245] described two patients who developed eczematous plaques following subcutaneous injections of heparin. Challenge tests with UH and LMWH were positive and histologic and immunohistochemical studies were consistent with a type IV hypersensitivity reaction. A subsequent investigation reported delayed-type hypersensitivity reactions in 7.5% of patients receiving subcutaneous LMWH injections; only one of these patients had anti−PF4 antibodies [246]. Patients with skin reactions are more likely to be women, have a body mass index >25, and a duration of heparin exposure >9 days. Pregnant patients on long-term heparin therapy are susceptible to delayed-type hypersensitivity reactions; 23% of the 177 women in two reports had skin involvement associated with LMWH therapy [247,248].

Cutaneous lesions complicating heparin therapy are either eczematous and pruritic, or take the form of a generalized, urticarial rash with angioedema; occasional patients have eosinophilia [249–251]. The risk of sensitization correlates with the molecular weight of the heparin and is greatest with UH, less with LMWH, and virtually absent with fondaparinux [252]. Patients allergic to one heparin (UH) often show cross-reactivity with others (LMWHs), so testing should be performed before starting a new product [253]. The offending agent should be discontinued; if urgent anticoagulation is required, switching from subcutaneous to intravenous heparin has been well-tolerated [254]. Fondaparinux or a DOAC might be appropriate for long-term therapy.

Necrotic skin lesions in association with heparin therapy were first described by Hall et al. [255] in 1980; they appeared at injection sites in their three patients after 6–11 days of heparin therapy. A later report of necrotic skin lesions with thrombosed vessels, distant from injection sites and accompanied by thrombocytopenia, shows that dermal necrosis is a feature of HIT [256]. A subsequent review of 21 patients with LMWH-associated skin necrosis observed thrombocytopenia ($<150,000/\mu L$) in 47% and HIT antibodies in 9 of 11 [257]. Warkentin considers necrotic skin lesions indicate HIT, but notes that only 10%–20% of patients with HIT antibodies develop these lesions [258]. Heparins should be discontinued immediately in such individuals and alternative anticoagulants prescribed.

A rare adverse effect of heparins is alopecia, reported with LMWH as well as UH. Dalteparin, given prior to hemodialysis sessions, induced hair loss in 5 women that subsided when regional citrate anticoagulation was substituted for the LMWH [259]. The alopecia in a 9-year-old girl receiving dalteparin for an internal jugular vein thrombosis also improved after discontinuing this anticoagulant [260]. Other authors observed alopecia after three weeks of enoxaparin therapy, with return to normal hair growth after the LMWH was withdrawn [261]. The alopecia is ascribed to drug-induced premature rest of the hair follicles (telogen effluvium) [262].

Osteoporosis

In 1965, osteoporosis and fractures were described in a 58 year old physician who was self-injecting UH, 15,000 U daily, to treat a myocardial infarction [263]. Eleven months later, a minor automobile accident resulted in multiple vertebral fractures, and subsequently, fractures occurred even with sneezing or lifting. Radiographs showed profound osteoporosis and he discontinued the heparin.

During long-term exposures to heparin, bone density declines and symptomatic fractures are reported in 2%–3% of patients [264]. Experiments in rats show that UH releases labeled calcium from osseous structures and promotes bone resorption by increasing osteoclast surface area; however,

LMWHs release calcium from bone only at concentrations well-above the therapeutic range [265,266]. In addition, UH and to a much lesser extent, LMWH [267], impair bone formation by decreasing osteoblast and osteoid surface. Studies using human osteoblast cell cultures reported significant inhibition of cell proliferation by four different LMWHs [268]. Heparin accumulates in bone, accounting for the lack of reversibility in loss of bone density [269].

Heparin administration for venous thromboembolism is often continued for months in pregnant patients. Significant decreases in proximal femoral bone density have been noted in a third of women given UH during pregnancy, but not in control pregnancies unexposed to heparin (P = 0.04), and differences from controls were still present 6 months postpartum (P = 0.03) [270]. Another study reported the results of dual photon absorptiometry of the lumbar spine in 50 women postpartum; the 25 women receiving long-term heparin therapy had 0.082 g/cm^2 lower bone density than the 25 untreated matched controls [271]. Substitution of UH with LMWH has reduced but not eliminated this complication of heparin therapy; on the other hand, osteoporosis does not occur with fondaparinux.

Hyperkalemia and elevation of transaminases

Heparin is an inhibitor of aldosterone production, resulting in increases in serum potassium above the normal range in 7% of exposed individuals [272]. Monitoring of potassium at 3−4 day intervals is recommended for patients with renal failure or other risk factors for hyperkalemia. Heparins also increase transaminases, but these changes do not appear to reflect hepatic or other organ impairment. Transaminase levels above baseline are observed in 95% of individuals receiving heparin therapy, are more common in men, and appear unrelated to the source of the anticoagulant [273,274]. The mean maximal fold-increase is 3.6 for alanine transaminase and 3.1 for aspartate transaminase; values return to the normal range when heparin exposure is terminated.

Summary

The incidence of major bleeding with unfractionated heparin is generally low; exceptions are in small, elderly women, individuals with platelet counts ≤ 50,000/μL, and patients with creatinine clearances ≤ 30 mL/minute. HIT occurs in 2%−3% of patients and can have devastating consequences if not recognized and managed appropriately. Other untoward effects of heparins are skin rashes, hypersensitivity reactions, and for patients on long-term therapy, osteoporosis. Low-molecular-weight heparins have similar bleeding risks but lower incidences of HIT, allergic reactions, and osteoporosis, and are more convenient to administer. Fondaparinux is rarely associated with

HIT, allergic reactions, or osteoporosis, but the lack of an antidote for bleeding is problematic because of the anticoagulant's relatively long half-life. The safe use of heparins requires familiarity with each agent and tailoring the dose and duration of treatment to address the needs and vulnerabilities of individual patients.

References

[1] Nieuwenhuis HK, Albada J, Banga JD, Sixma JJ. Identification of risk factors for bleeding during treatment of acute venous thromboembolism with heparin or low molecular weight heparin. Blood 1991;78:2337−43.

[2] Morabia A. Heparin doses and major bleedings. The Lancet 1986;i:1278−9.

[3] Kearon C, Akl EA, Ornelas J, Blaivas A, Jimenez D, Bounameaux H, et al. Antithrombotic therapy for VTE disease: CHEST Guideline and Expert Panel Report. Chest 2016;149:315−52.

[4] Kraaijpoel N, van Es N, Bleker SM, Brekelmans MPA, Eerenberg ES, Middeldorp S, et al. Clinical impact and course of anticoagulant-related major bleeding in cancer patients. Thromb Haemost 2018;118:174−81.

[5] Donato J, Campigotto F, Uhlmann EJ, Coletti E, Neuberg D, Weber GM, et al. Intracranial hemorrhage in patients with brain metastases treated with therapeutic enoxaparin: a matched cohort study. Blood 2015;126:494−9.

[6] Jick H, Slone D, Borda IT, Shapiro S. Efficacy and toxicity of heparin in relation to age and sex. N Engl J Med 1968;279:284−6.

[7] Jick H, Slone D, Borda IT, Shapiro S. Heparin therapy. N Engl J Med 1968;279:887−8.

[8] Campbell NRC, Hull RD, Brant R, Hogan DB, Pineo GF, Raskob GE. Aging and heparin-related bleeding. Arch Intern Med 1996;156:857−60.

[9] Campbell NR, Hull RD, Brant R, Hogan DB, Pineo GF, Raskob GE. Different effects of heparin in males and females. Clin Invest Med 1998;21:71−8.

[10] Jick H, Porter J. Drug-induced gastrointestinal bleeding. The Lancet 1978;ii:87−9.

[11] Mant MJ, O'Brien BD, Thong KL, Hammond GW, Birtwhistle RV, Grace MG. Haemorrhagic complications of heparin therapy. The Lancet 1977;i:1133−5.

[12] Hull RD, Raskob GE, Pineo GF, Green D, Trowbridge AA, Elliott CG, et al. Subcutaneous low-molecular-weight heparin compared with continuous intravenous heparin in the treatment of proximal-vein thrombosis. N Engl J Med 1992;326:975−82.

[13] Hommes DW, Bura A, Mazzolai L, Buller HR, ten Cate JW. Subcutaneous heparin compared with continuous intravenous heparin administration in the initial treatment of deep vein thrombosis: a meta-analysis. Ann Intern Med 1992;116:279−84.

[14] Quinlan DJ, McQuillan A, Eikelboom JW. Low-molecular-weight heparin compared with intravenous unfractionated heparin for treatment of pulmonary embolism: a meta-analysis of randomized, controlled trials. Ann Intern Med 2004;140:175−83.

[15] Bucci C, Geerts WH, Sinclair A, Fremes SE. Comparison of the effectiveness and safety of low-molecular weight heparin versus unfractionated heparin anticoagulation after heart valve surgery. Am J Cardiol 2011;107:591−4.

[16] Gould MK, Dembitzer AD, Doyle RL, Hastie TJ, Garber AM. Low-molecular weight heparins compared with unfractionated heparin for treatment of acute deep venous thrombosis. Ann Intern Med 1999;130:800−9

[17] Kearon C, Ginberg JS, Julian JA, Douketis J, Solymoss S, Ockelford P, et al. A comparison of fixed-dose weight-adjusted unfractionated heparin and low-molecular-weight heparin for acute treatment of venous thromboembolism. JAMA 2006;296:935–42.

[18] Buller HR, Davidson BL, Decousus H, Gallus A, Gent M, Piovella F, et al. Fondaparinux or enoxaparin for the initial treatment of symptomatic deep venous thrombosis. A randomized trial. Ann Intern Med 2004;140:867–73.

[19] Imperiale TF, Speroff T. A meta-analysis of methods to prevent venous thromboembolism following total hip replacement. JAMA 1994;271:1780–5.

[20] Nurmohamed MT, Verhaeghe R, Haas S, Iriate JA, Vogel G, van Rij AM, et al. A comparative trial of a low molecular weight heparin (enoxaparin) versus standard heparin for the prophylaxis of postoperative deep vein thrombosis in general surgery. Am J Surg 1995;169:567–71.

[21] McLeod RS, Geerts WH, Sniderman KW, Greenwood C, Gregoire RC, Taylor BM, et al. Subcutaneous heparin versus low-molecular-weight heparin as thromboprophylaxis in patients undergoing colorectal surgery: results of the Canadian colorectal DVT prophylaxis trial: a randomized, double-blind trial. Ann Surg 2001;233:438–44.

[22] ENOXACAN Study Group. Efficacy and safety of enoxaparin versus unfractionated heparin for prevention of deep vein thrombosis in elective cancer surgery: a double-blind randomized multicenter trial with venographic assessment. Br J Surg 1997;84:1099–103.

[23] Kakkar VV, Cohen AT, Edmonson RA, Phillips MJ, Cooper DJ, Das SK, et al. Low molecular weight versus standard heparin for prevention of venous thrombosis after major abdominal surgery. The Lancet 1993;341:259–65.

[24] Gazzaniga GM, Angelini G, Pastorino G, Santoro E, Lucchini M, Dal Prà ML. Enoxaparin in the prevention of deep venous thrombosis after major surgery: multicentric study. The Italian Study Group. Int Surg 1993;78:271–5.

[25] Colwell Jr CW, Spiro TE, Trowbridge AA, Morris BA, Kwaan HC, Blaha JD, et al. Use of enoxaparin, a low-molecular-weight heparin, and unfractionated heparin for the prevention of deep venous thrombosis after elective hip replacement. JBJS 1994;76-A:3–14.

[26] Francis CW, Pellegrini Jr VD, Totterman S, Boyd Jr AD, Marder VJ, Liebert KM, et al. Prevention of deep-vein thrombosis after total hip arthroplasty. JBJS 1997;79-A:1365–72.

[27] Hull RD, Raskob G, Pineo G, Rosenbloom D, Evans W, Mallory T, et al. A comparison of subcutaneous low-molecular weight heparin with warfarin sodium for prophylaxis against deep-vein thrombosis after hip or knee implantation. N Engl J Med 1993;329:1370–6.

[28] Thomas DP. Does low molecular weight heparin cause less bleeding? Thromb Haemost 1997;78:1422–5.

[29] Bounameaux H, Perneger T. Fondaparinux: a new synthetic pentasaccharide for thrombosis prevention. The Lancet 2002;359:1710–11.

[30] Lu E, Shatzel JJ, Salati J, DeLoughery TG. The safety of low-molecular-weight heparin during and after pregnancy. Obstet Gynecol Surv 2017;72:721–9.

[31] Carrier M, Khorana AA, Zwicker JI, Noble S, Lee AYY, on behalf of the Subcommittee on Haemostasis and Malignancy of the SSC of the ISTH. Management of challenging cases of patients with cancer-associated thrombosis including recurrent thrombosis and bleeding: guidance from the SSC of the ISTH. J Thromb Haemost 2013;11760–5.

[32] LaPointe NM, Chen AY, Alexander KP, Roe MT, Pollack Jr CV, Lytle BL, et al. Enoxaparin dosing and associated risk of in-hospital bleeding death in patients with non-ST-segment elevation acute coronary syndromes. Arch Intern Med 2007;167:1539–44.

[33] Lim W, Dentali F, Eikelboom JW, Crowther MA. Meta-analysis: low-molecular-weight heparin and bleeding in patients with severe renal insufficiency. Ann Intern Med 2006;144:673—84.

[34] Willbanks OL, Fuller CH. Femoral neuropathy due to retroperitoneal bleeding. Arch Intern Med 1973;132:83—6.

[35] Wong JS. Spontaneous suprachoroidal haemorrhage in a patient receiving low-molecular weight heparin (fraxiparine) therapy. Aust N Z J Ophthalmol 1999;27:433—4.

[36] Amador E. Adrenal hemorrhage during anticoagulant therapy. Ann Intern Med 1965;63:559—71.

[37] Horlocker TT. Complications of spinal and epidural anesthesia. Anesthesiol Clin North Am 2000;18:461—85.

[38] Horlocker TT. Regional anaesthesia in the patient receiving antithrombotic and antiplatelet therapy. Br J Anaesth 2011;107(Suppl 1):i96—i106.

[39] Horlocker TT, Vandermeuelen E, Kopp SL, Gogarten W, Leffert LR, Benzon HT. Regional anesthesia in the patient receiving antithrombotic or thrombolytic therapy: American Society of Regional Anesthesia and Pain Medicine Evidence-Based Guidelines (Fourth Edition). Reg Anesth Pain Med 2018;43:263—309.

[40] Douketis JD, Kinnon K, Crowther MA. Anticoagulant effect at the time of epidural catheter removal in patients receiving twice-daily or once-daily low-molecular-weight heparin and continuous epidural analgesia after orthopedic surgery. Thromb Haemost 2002;88:37—40.

[41] Green D, Klement P, Liao P, Weitz J. Interaction of low molecular weight heparin with ketorolac. J Lab Clin Med 1996;127:583—7.

[42] Greer IA, Gibson JL, Young A, Johnstone J, Walker ID. Effect of ketorolac and low-molecular-weight heparin individually and in combination on haemostasis. Blood Coagul Fibrinolysis 1999;10:367—73.

[43] Singelyn FJ, Verheyen CC, Piovella F, Van Aken HK, Rosencher N, EXPERT Study Investigators. The safety and efficacy of extended thromboprophylaxis with fondaparinux after major orthopedic surgery of the lower limb with or without a neuraxial or deep peripheral nerve catheter: the EXPERT Study. Anesth Analg 2007;105:1540—7.

[44] Piran S, Schulman S. Treatment of bleeding complications in patients on anticoagulant therapy. Blood 2019;133:425—35.

[45] Wolzt M, Weltermann A, Nieszpaur-Los M, Schneider B, Fassolt A, Lechner K, et al. Studies on the neutralizing effects of protamine on unfractionated and low molecular weight heparin (Fragmin®) at the site of activation of the coagulation system in man. Thromb Haemost 1995;73:439—43.

[46] Monagle P, Michelson AD, Bovill E, Andrew M. Antithrombotic therapy in children. Chest 2001;119(1 Suppl):344S—70S.

[47] Levy JH, Schwieger IM, Zaidan JR, Faraj BA, Weintraub WS. Evaluation of patients at risk for protamine reactions. J Thorac Cardiovasc Surg 1989;98:200—4.

[48] Massonnet-Castel S, Pelissier E, Bara L, Terrier E, Abry B, Guibourt P, et al. Partial reversal of low molecular weight heparin (PK10169) anti-Xa activity by protamine sulfate: in vitro and in vivo study during cardiac surgery with extracorporeal circulation. Haemostasis 1986;16:139—46.

[49] Van Veen JJ, Maclean RM, Hampton KK, Laidlaw S, Kitchen S, Toth P, et al. Protamine reversal of low molecular weight heparin: clinically effective? Blood Coagul Fibrinolysis 2011;22(7).565—70.

[50] Hu Q, Brady JO. Recombinant activated factor VII for treatment of enoxaparin-induced bleeding. Mayo Clin Proc 2004;79:827.

[51] Crowther M, Levy G, Lu G, Conley PB, Castillo J, Hollenbach S, et al. Reversal of enoxaparin-induced anticoagulation in healthy subjects by andexanet alfa (PRT064445), an antidote for direct and indirect fXa inhibitors-a phase 2 randomized, double-blind, placebo-controlled trial. J Thromb Hemost 2014;12(Suppl.1) COA01, p.7.

[52] Connolly SJ, Crowther M, Eikelboom JW, Gibson CM, Curnutte JT, Lawrence JH, et al. Full study report of andexanet alfa for bleeding associated with factor Xa inhibitors. N Engl J Med 2019;380:1326−35.

[53] Bijsterveld NR, Moons AH, Boekholdt SM, van Aken BE, Fennema H, Peters RJ, et al. Ability of recombinant factor VIIa to reverse the anticoagulant effect of the pentasaccharide fondaparinux in healthy volunteers. Circulation 2002;106:2550−4.

[54] Luporsi P, Chopard R, Janin S, Racadot E, Bernard Y, Ecarnot F, et al. Use of recombinant factor VIIa (NovoSeven®) in 8 patients with ongoing life-threatening bleeding treated with fondaparinux. Acute Card Care 2011;13:93−8.

[55] Desmurs-Clavel H, Huchon C, Chatard B, Negrier C, Dargaud Y. Reversal of the inhibitory effect of fondaparinux on thrombin generation by rFVIIa, aPCC and PCC. Thromb Res 2009;123:796−8.

[56] Frontera JA, Lewin 3rd JJ, Rabinstein AA, Aisiku IP, Alexandrov AW, Cook AM, et al. Guideline for reversal of antithrombotics in intracranial hemorrhage: a statement for healthcare professionals from the neurocritical care society and society of critical care medicine. Neurocrit Care 2016;24:6−46.

[57] Nieto JA, Bruscas MJ, Ruiz-Ribo D, Trujillo-Santos J, Valle R, Ruiz-Gimenez N, et al. Acute venous thromboembolism in patients with recent major bleeding. The influence of the site of bleeding and the time elapsed on outcome. J Thromb Haemost 2006;4:2367−72.

[58] Sengupta N, Feuerstein JD, Patwardhan VR, Tapper EB, Ketwaroo GA, Thaker AM, et al. The risks of thromboembolism vs. recurrent gastrointestinal bleeding after interruption of systemic anticoagulation in hospitalized inpatients with gastrointestinal bleeding: a prospective study. Am J Gastroenterol 2015;110:328−35.

[59] Witt DM. What to do after the bleed: resuming anticoagulation after major bleeding. Hematol/Am Soc Hematol Edu Program 2016;2016:620−4.

[60] Natelson EA, Lynch EC, Alfrey Jr CP, Gross JB. Heparin-induced thrombocytopenia. Ann Intern Med 1969;71:1121−5.

[61] Kelton JG, Warkentin TE. Heparin-induced thrombocytopenia: a historical perspective. Blood 2008;112:2607−15.

[62] Rhodes GR, Dixon RH, Silver D. Heparin induced thrombocytopenia with thrombotic and hemorrhagic manifestations. Surg Gynecol Obstet 1973;136:409−16.

[63] Rhodes GR, Dixon RH, Silver D. Heparin induced thrombocytopenia: eight cases with thrombotic-hemorrhagic complications. Ann Surg 1977;186:752−8.

[64] Hussey CV, Bernhard VM, McLean MR, Fobian JE. Heparin induced platelet aggregation: in vitro confirmation of thrombotic complications. Associated with heparin therapy. Ann Clin Lab Sci 1979;9:487−93.

[65] Babcock RB, Dumper CW, Scharfman WB. Heparin-induced immune thrombocytopenia. N Engl J Med 1976;295:237−41.

[66] Green D, Harris K, Reynolds N, Roberts M, Patterson R. Heparin immune thrombocytopenia: evidence for a heparin-platelet complex as the antigenic determinant. J Lab Clin Med 1978;91:167−75.

[67] Amiral J, Bridey F, Dreyfus M, Vissoc AM, Fressinaud E, Wolf M, et al. Platelet factor 4 complexed to heparin is the target for antibodies generated in heparin-induced thrombocytopenia. Thromb Haemost 1992;68:95—6.

[68] Horne MK, Alkins BR. Importance of PF4 in heparin-induced thrombocytopenia: confirmation with gray platelets. Blood 1995;85:1408—9.

[69] Cines DB, Tomaski A, Tannenbaum S. Immune endothelial-cell injury in heparin-associated thrombocytopenia. N Engl J Med 1987;316:581—9.

[70] Warkentin TE, Hayward CPM, Boshkov LK, Santos AV, Sheppard J-AI, Bode AP, et al. Sera from patients with heparin-induced thrombocytopenia generate platelet-derived microparticles with procoagulant activity: an explanation for the thrombotic complications of heparin-induced thrombocytopenia. Blood 1994;84:3691—9.

[71] Kelton JG, Hirsh J. Bleeding associated with antithrombotic therapy. Semin Hematol 1980;17:259—91.

[72] Eika C, Godal HC, Laake K, Hamborg T. Low incidence of thrombocytopenia during treatment with hog mucosa and beef lung heparin. Scand J Haematol 1980;25:19—24.

[73] Ansell J, Slepchuk Jr N, Kumar R, Lopez A, Southard L, Deykin D. Heparin induced thrombocytopenia: a prospective study. Thromb Haemost 1980;43:61—5.

[74] Green D, Martin GJ, Shoichet SH, DeBacker N, Bomaleski JS, Lind RN. Thrombocytopenia in a prospective, randomized, double-blind trial of bovine and porcine heparin. Am J Med Sci 1984;288:60—4.

[75] Bell WR, Royall RM. Heparin-associated thrombocytopenia: a comparison of three heparin preparations. N Engl J Med 1980;303:902—7.

[76] Konkle BA, Bauer TL, Arepally G, Cines DB, Poncz M, McNulty S, et al. Heparin-induced thrombocytopenia: bovine versus porcine heparin in cardiopulmonary bypass surgery. Ann Thorac Surg 2001;71:1920—4.

[77] Stead RB, Schafer AI, Rosenberg RD, Handin RI, Josa M, Khuri SF. Heterogeneity of heparin lots associated with thrombocytopenia and thromboembolism. Am J Med 1984;77:185—8.

[78] Kato S, Takahashi K, Ayabe K, Samad R, Fukaya E, Friedmann P, et al. Heparin-induced thrombocytopenia: analysis of risk factors in medical inpatients. Br J Haematol 2011;154 (3):373—7.

[79] Warkentin TE, Kelton JG. Temporal aspects of heparin-induced thrombocytopenia. N Engl J Med 2001;344:1286—92.

[80] Potzsch B, Klovekorn W-P, Madlener K. Use of heparin during cardiopulmonary bypass in patients with a history of heparin-induced thrombocytopenia. N Engl J Med 2000;343:515.

[81] Warkentin TE, Sheppard J-AI. Serological investigation of patients with a previous history of heparin-induced thrombocytopenia who are re-exposed to heparin. Blood 2014;123:2485—93.

[82] Selleng S, Haneya A, Hirt S, Selleng K, Schmid C, Greinacher A. Management of anticoagulation in patients with subacute heparin-induced thrombocytopenia scheduled for heart transplantation. Blood 2008;112:4024—7.

[83] Warkentin TE, Levine MN, Hirsh J, Horsewood P, Roberts RS, Gent M, et al. Heparin-induced thrombocytopenia in patients treated with low molecular-weight heparin or unfractionated heparin. N Engl J Med 1995;332:1330—5.

[84] Martel N, Lee J, Wells PS. Risk for heparin-induced thrombocytopenia with unfractionated and low-molecular-weight heparin thromboprophylaxis: a meta-analysis. Blood 2005;106:2710—15.

[85] Salter BS, Weiner MM, Trinh MA, Heller J, Evans AS, Adams DH, et al. Heparin-induced thrombocytopenia: a comprehensive clinical review. J Am Coll Cardiol 2016;67:2519–32.

[86] Warkentin TE, Lim W. Can heparin-induced thrombocytopenia be associated with fonda-parinux use? Reply to a rebuttal. J Thromb Haemost 2008;6:1243–6.

[87] Warkentin TE, Sheppard JA, Horsewood P, Simpson PJ, Moore JC, Kelton JG. Impact of the patient population on the risk for heparin-induced thrombocytopenia. Blood 2000;96:1703–8.

[88] Warkentin TE, Cook RJ, Marder VJ, Sheppard JA, Moore JC, Eriksson BI, et al. Anti-platelet factor 4/heparin antibodies in orthopedic surgery patients receiving antithrombo-tic prophylaxis with fondaparinux or enoxaparin. Blood 2005;106:3791–6.

[89] Savi P, Chong BH, Greinacher A, Gruel Y, Kelton JG, Warkentin TE, et al. Effect of fondaparinux on platelet activation in the presence of heparin-dependent antibodies: a blinded comparative multicenter study with unfractionated heparin. Blood 2005;105:139–44.

[90] Chong BH, Pilgrim RL, Cooley MA, Chesterman CN. Increased expression of platelet IgG Fc receptors in immune heparin-induced thrombocytopenia. Blood 1993;81:988–93.

[91] Reilly MP, Taylor SM, Hartman NK, Arepally GM, Sachias BS, Cines DB, et al. Heparin-induced thrombocytopenia/thrombosis in a transgenic mouse model requires human platelet factor 4 and platelet activation through Fcγ RIIA. Blood 2001;98:2442–7.

[92] Carlsson LE, Santoso S, Baurichter G, Kroll H, Papenberg S, Eichler P, et al. Heparin-induced thrombocytopenia: new insights into the impact of the FcγRIIa-R-H131 poly-morphism. Blood 1998;92:1526–31.

[93] Rollin J, Pouplard C, Gruel Y. Risk factors for heparin-induced thrombocytopenia: focus on Fcγ receptors. Thromb Haemost 2016;116:799–805.

[94] Matzinger P. Tolerance, danger, and the extended family. Annu Rev Immunol 1994;12:991–1045.

[95] Lubenow N, Hinz P, Thomaschewski S, Lietz T, Vogler M, Ladwig A, et al. The sever-ity of trauma determines the immune response to PF4/heparin and the frequency of heparin-induced thrombocytopenia. Blood 2010;115:1797–803.

[96] Girolami B, Prandoni P, Stefani PM, Tanduo C, Sabbion P, Eichler P, et al. The inci-dence of heparin-induced thrombocytopenia in hospitalized medical patients treated with subcutaneous unfractionated heparin: a prospective cohort study. Blood 2003;101:2955–9.

[97] Klinkhammer B, Gruchalla M. Is there an association between heparin-induced thrombo-cytopenia (HIT) and autoimmune disease? WMJ 2018;117:13–17.

[98] Pauzner R, Greinacher A, Selleng K, Althaus K, Shenkman B, Seligsohn U. False-positive tests for heparin-induced thrombocytopenia in patients with antiphospholipid syndrome and systemic lupus erythematosus. J Thromb Haemost 2009;7:1070–4.

[99] Asmis LM, Segal JB, Plantinga LC, Fink NE, Kerman JS, Kickler TS, et al. Heparin-induced antibodies and cardiovascular risk in patients on dialysis. Thromb Haemost 2008;100:498–504.

[100] Matsuo T, Kobayashi H, Matsuo M, Wanaka K, Nakamoto H, Matsushima H, et al. Frequency of anti-heparin–PF4 complex antibodies (HIT antibodies) in uremic patients on chronic intermittent hemodialysis. Pathophysiol Haemost Thromb 2006;35:445–50.

[101] Warkentin TE. Heparin-induced thrombocytopenia in critically ill patients. Semin Thromb Hemost 2015;41:49–60.

[102]   Warkentin TE, Sheppard J-AI, Sigouin CS, Kohlmann T, Eichler P, Greinacher A. Gender imbalance and risk factor interactions in heparin-induced thrombocytopenia. Blood 2006;108:2937−41.

[103]   Salzman EW, Rosenberg RD, Smith MH, Lindon JN, Favreau L. Effect of heparin and heparin fractions on platelet aggregation. J Clin Invest 1980;65:64−73.

[104]   Greinacher A, Potzsch B, Amiral J, Dummel V, Eichner A, Mueller-Eckhardt C. Heparin-associated thrombocytopenia: isolation of the antibody and characterization of a multimolecular PF4-heparin complex as the major antigen. Thromb Haemost 1994;71:247−51.

[105]   Rauova L, Poncz M, McKenzie SE, Reilly MP, Arepally G, Weisel JW, et al. Ultralarge complexes of PF4 and heparin are central to the pathogenesis of heparin-induced thrombocytopenia. Blood 2005;105:131−8.

[106]   Kelton JG, Sheridan D, Santos A, Smith J, Steeves K, Smith C, et al. Heparin-induced thrombocytopenia: laboratory studies. Blood 1988;72:925−30.

[107]   Adelman B, Sobel M, Fujimura Y, Ruggeri ZM, Zimmerman TS. Heparin-associated thrombocytopenia: observations on the mechanism of platelet aggregation. J Lab Clin Med 1989;113:204−10.

[108]   Chong BH. Heparin-induced thrombocytopenia. Br J Hematol 1995;89:431−9.

[109]   Rauova L, Hirsch JD, Greene TK, Zhai L, Hayes VM, Kowalska MA, et al. Monocyte-bound PF4 in the pathogenesis of heparin-induced thrombocytopenia. Blood 2010;116:5021−31.

[110]   Towne JB, Bernhard VM, Hussey C, Garancis JC. White clot syndrome. Peripheral vascular complications of heparin therapy. Arch Surg 1979;114:372−7.

[111]   Hrushesky WJ. Subcutaneous heparin-induced thrombocytopenia. Arch Intern Med 1978;138:1489−91.

[112]   Hollander G, Bashevkin M, Feldman H, Lichstein E. Heparin-induced thrombotic thrombocytopenia. Chest 1981;79:234−6.

[113]   Cimo PL, Moake JL, Weinger RS, Ben-Menachem Y, Khalil KG. Heparin-induced thrombocytopenia: association with a platelet aggregating factor and arterial thromboses. Am J Hem 1979;6:125−33.

[114]   Cines DB, Kaywin P, Bina M, Tomaski A, Schreiber AD. Heparin-associated thrombocytopenia. N Engl J Med 1980;303:788−95.

[115]   Sachais BS, Litvinov RI, Yarovoi SV, Rauova L, Hinds JL, Rux AH, et al. Dynamic antibody-binding properties in the pathogenesis of HIT. Blood 2012;120:1137−42.

[116]   Greinacher A, Krauel K, Jensch I. HIT-antibodies promote their own antigen. Blood 2012;120:930−1.

[117]   Greinacher A, Gopinadhan M, Gunther J-U, Omer-Adam MA, Strobel U, Warkentin TE, et al. Close approximation of two platelet factor 4 tetramers by charge neutralization forms the antigens recognized by HIT antibodies. Arterioscler Thromb Vasc Biol 2006;26:2386−93.

[118]   Newman PM, Chong BH. Heparin-induced thrombocytopenia: new evidence for the dynamic binding of purified anti−PF4-heparin antibodies to platelets and the resultant platelet activation. Blood 2000;96:182−7.

[119]   Bacsi S, DePalma R, Visentin GP, Gorski J, Aster RH. Complexes of heparin and platelet factor 4 specifically stimulate T cells from patients with heparin-induced thrombocytopenia/thrombosis. Blood 1999;94:208−15.

[120]   Suvarna S, Rauova L, McCracken EKE, Goss CM, Sachais BS, McKenzie SE, et al. PF4/heparin complexes are T cell-dependent antigens. Blood 2005;106:929−31.

[121] Zheng Y, Yu M, Podd A, Yuan L, Newman DK, Wen R, et al. Critical role for mouse marginal zone B cells in PF4/heparin antibody production. Blood 2013;121:3484−92.

[122] Zheng Y, Wang AW, Yu M, Padmanabhan A, Tourdot BE, Newman DK, et al. B-cell tolerance regulates production of antibodies causing heparin-induced thrombocytopenia. Blood 2014;123:931−4.

[123] Warkentin TE, Sheppard JI, Moore JC, Cook RJ, Kelton JG. Studies of the immune response in heparin-induced thrombocytopenia. Blood 2009;113:4963−9.

[124] Greinacher A, Kohlmann T, Strobel U, Sheppard JA, Warkentin TE. The temporal profile of the anti−PF4/heparin immune response. Blood 2009;113:4970−6.

[125] Harenberg J, Wang LC, Hoffmann U, Huhle G, Feuring M. Laboratory diagnosis of heparin-induced thrombocytopenia type II after clearance of platelet factor 4/heparin complex. J Lab Clin Med 2001;137:408−13.

[126] Kelton JG, Smith JW, Warkentin TE, Hayward CPM, Denomme GA, Horsewood P. Immunoglobulin G from patients with heparin-induced thrombocytopenia binds to a complex of heparin and platelet factor 4. Blood 1994;83:3232−9.

[127] Horne MK, Alkins BR. Platelet binding of IgG from patients with heparin-induced thrombocytopenia. J Lab Clin Med 1996;127:435−42.

[128] Rollin J, Pouplard C, Sung HC, Leroux D, Saada A, Gouilleux-Gruart V, et al. Increased risk of thrombosis in FcγRIIA 131RR patients with HIT due to defective control of platelet activation by plasma IgG2. Blood 2015;125:2397−404.

[129] Yeung J, Tourdot BE, Fernandez-Perez P, Vesci J, Ren J, Smyrniotis CJ, et al. Platelet 12-LOX is essential for FcγRIIa-mediated platelet activation. Blood 2014;124:2271−9.

[130] Chong BH, Pitney WR, Castaldi PA. Heparin-induced thrombocytopenia: association of thrombotic complications with heparin-dependent IgG antibody that induces thromboxane synthesis and platelet aggregation. The Lancet 1982;ii:1246−9.

[131] Hughes M, Hayward CPM, Warkentin TE, Horsewood P, Chorneyko KA, Kelton JG. Morphological analysis of microparticle generation in heparin-induced thrombocytopenia. Blood 2000;96:188−94.

[132] Munnix IC, Cosemans JM, Auger JM, Heemskerk JW. Platelet response heterogeneity in thrombus formation. Thromb Haemost 2009;102:1149−56.

[133] Tardy-Poncet B, Piot M, Chapelle C, France G, Campos L, Garraud O, et al. Thrombin generation and heparin-induced thrombocytopenia. J Thromb Haemost 2009;7:1474−81.

[134] Visentin GP, Moghaddan M, Beery SH, McFarland JG, Aster RH. Heparin is not required for detection of antibodies associated with heparin-induced thrombocytopenia/thrombosis. J Lab Clin Med 2001;138:22−31.

[135] Padmanabhan A, Jones CG, Bougie DW, Curtis BR, McFarland JG, Wang D, et al. Heparin-independent, PF4-dependent binding of HIT antibodies to platelets: implications for HIT pathogenesis. Blood 2015;125:155−61.

[136] Pouplard C, Lochmann S, Renard B, Herault O, Colombat P, Amiral J, et al. Induction of monocyte tissue factor expression by antibodies to heparin-platelet factor 4 complexes developed in heparin-induced thrombocytopenia. Blood 2001;97:3300−2.

[137] Kasthuri RS, Glover SL, Jonas W, McEachron T, Pawlinski R, Arepally GM, et al. PF4/heparin-antibody complex induces monocyte tissue factor expression and release of tissue factor positive microparticles by activation of FcγR1. Blood 2012;119:5285−93.

[138] Tutwiler V, Madeeva D, Ahn HS, Andrianova I, Hayes V, Zheng XL, et al. Platelet transactivation by monocytes promotes thrombosis in heparin-induced thrombocytopenia. Blood 2016;127:464−72.

[139] Visentin GP, Ford SE, Scott JP, Aster RH. Antibodies from patients with heparin-induced thrombocytopenia/thrombosis are specific for platelet factor 4 complexed with heparin or bound to endothelial cells. J Clin Invest 1994;93:81−8.

[140] Dai J, Madeeva D, Hayes V, Ahn HS, Tutwiler V, Arepally GM, et al. Dynamic intercellular redistribution of HIT antigen modulates heparin-induced thrombocytopenia. Blood 2018;132:727−34.

[141] Rauova L, Zhai L, Kowalska MA, Arepally GM, Cines DB, Poncz M. Role of platelet surface antigenic complexes in heparin-induced thrombocytopenia pathogenesis: diagnostic and therapeutic implications. Blood 2006;107:2346−53.

[142] Matsuo T, Suzuki S, Matsuo M, Kobayasi H. Preexisting antibodies to platelet factor 4-heparin complexes in patients with acute coronary syndrome who have no history of heparin exposure. Pathophysiol Haemost Thromb 2005;34:18−22.

[143] Flego D, Liuzzo G, Weyand CM, Crea F. Adaptive immunity dysregulation in acute coronary syndromes: from cellular and molecular basis to clinical implications. J Am Coll Cardiol 2016;68:2107−17.

[144] Dudek AZ, Pennell CA, Decker TD, Young TA, Key NS, Slungaard A. Platelet factor 4 binds to glycanated forms of thrombomodulin and to protein C. A potential mechanism for enhancing generation of activated protein C. J Biol Chem 1997;272:31785−92.

[145] Kowalska MA, Krishnaswamy S, Rauova L, Zhai L, Hayes V, Amirikian K, et al. Antibodies associated with heparin-induced thrombocytopenia (HIT) inhibit activated protein C generation: new insights into the prothrombotic nature of HIT. Blood 2011;118:2882−8.

[146] Khairy M, Lasne D, Brohard-Bohn B, Aiach M, Rendu F, Bachelot-Loza C. A new approach in the study of the molecular and cellular events implicated in heparin-Induced thrombocytopenia. Thromb Haemost 2001;85:1090−6.

[147] Xiao Z, Visentin GP, Dayananda KM, Neelamegham S. Immune complexes formed following the binding of anti-platelet factor 4 (CXCL4) antibodies to CXCL4 stimulate human neutrophil activation and cell adhesion. Blood 2008;112:1091−100.

[148] Perdomo J, Leung HHL, Ahmadi Z, Yan F, Chong JJH, Passam FH, et al. Neutrophil activation and NETosis are the major drivers of thrombosis in heparin-induced thrombocytopenia. Nat Commun 2019;10:1322.

[149] Gollomp K, Kim M, Johnston I, Hayes V, Welsh J, Arepally GM, et al. Neutrophil accumulation and NET release contribute to thrombosis in HIT. JCI Insight 2018;3 pii: 99445.

[150] Baggiolini M, Clark-Lewis I. Interleukin-8, a chemotactic and inflammatory cytokine. FEBS Lett 1992;307:97−101.

[151] Amiral J, Marfaing-Koka A, Wolf M, Alessi MC, Tardy B, Boyer-Neumann C, et al. Presence of autoantibodies to interleukin-8 or neutrophil-activating peptide-2 in patients with heparin-associated thrombocytopenia. Blood 1996;88:410−16.

[152] Arepally GM, Mayer IM. Antibodies from patients with heparin-induced thrombocytopenia stimulate monocytic cells to express tissue factor and secrete interleukin-8. Blood 2001;98:1252−4.

[153] Prechel MM, Walenga JM. Complexes of platelet factor 4 and heparin activate Toll-like receptor 4. J Thromb Haemost 2015;13:665−70.

[154] Krauel K, Weber C, Brandt S, Zähringer U, Mamat U, Greinacher A, et al. Platelet factor 4 binding to lipid A of Gram-negative bacteria exposes PF4/heparin-like epitopes. Blood 2012;120:3345 52.

[155] Greinacher A, Holtfreter B, Krauel K, Gätke D, Weber C, Ittermann T, et al. Association of natural anti-platelet factor 4/heparin antibodies with periodontal disease. Blood 2011;118:1395−401.

[156] Gruel Y, Watier H. Bacteria and HIT: a close connection? Blood 2011;1171105−6.

[157] Lande R, Lee EY, Palazzo R, Marinari B, Pietraforte I, Santos GS, et al. CXCL4 assembles DNA into liquid crystalline complexes to amplify TLR9-mediated interferon-α production in systemic sclerosis. Nat Commun 2019;10(1):1731.

[158] Rauova L, Arepally GM, Ciines DB, Poncz M. Chapter 9: cellular and molecular immunoopathogenesis of heparin-induced thrombocytopenia. In: Warkentin TE, Greinacher A, editors. Heparin-induced thrombocytopenia. 5th ed. Boca Raton: CRC Press; 2013. p. 243−4.

[159] Selleng S, Selleng K, Wollert H-G, Muellejans B, Lietz T, Warkentin TE, et al. Heparin-induced thrombocytopenia in patients requiring prolonged intensive care unit treatment after cardiopulmonary bypass. J Thromb Haemost 2008;6:428−35.

[160] Selleng S, Malowsky B, Strobel U, Wessel A, Ittermann T, Wollert H-G, et al. Early-onset and persisting thrombocytopenia in post-cardiac surgery patients is rearely due to heparin-induced thrombocytopenia, even when antibody tests are positive. J Thromb Haemost 2010;8:30−6.

[161] Rice L, Attisha WK, Drexler A, Francis JL. Delayed-onset heparin-induced thrombocytopenia. Ann Intern Med 2002;136:210−15.

[162] Warkentin TE, Kelton JG. Delayed-onset heparin-induced thrombocytopenia and thrombosis. Ann Intern Med 2001;135:502−6.

[163] Phelan BK. Heparin-associated thrombosis without thrombocytopenia. Ann Intern Med 1983;99:637−8.

[164] Hach-Wunderle V, Kainer K, Krug B, Muller-Berghaus G, Potzsch B. Heparin-associated thrombosis despite normal platelet counts. The Lancet 1994;344:469−70.

[165] Warkentin TE, Basciano PA, Knopman J, Bernstein RA. Spontaneous heparin-induced thrombocytopenia syndrome: 2 new cases and a proposal for defining this disorder. Blood 2014;123:3651−4.

[166] Jaax ME, Krauel K, Marschall T, Brandt S, Gansler J, Fürll B, et al. Complex formation with nucleic acids and aptamers alters the antigenic properties of platelet factor 4. Blood 2013;122:272−81.

[167] Chong BH, Chong J-H. HIT: nucleic acid masquerading as heparin. Blood 2013;122:156−8.

[168] Greinacher A. Me or not me? The danger of spontaneity. Blood 2014;123:3536−8.

[169] Pohl C, Klockgether T, Greinacher A, Hanfland P, Harbrecht U. Neurological complications in heparin-induced thrombocytopenia. The Lancet 1999;353:1678−9.

[170] Hong AP, Cook DJ, Sigouin CS, Warkentin TE. Central venous catheters and upper-extremity deep-vein thrombosis complicating immune heparin-induced thrombocytopenia. Blood 2003;101:3049−51.

[171] Atkinson JLD, Sundt Jr TM, Kazmier FJ, Bowie EJW, Whisnant JP. Heparin-induced thrombocytopenia and thrombosis in ischemic stroke. Mayo Clin Proc 1988;63:353−61.

[172] Warkentin TE, Safyan EL, Linkins L-A. Heparin-induced thrombocytopenia presenting as bilateral adrenal hemorrhages. N Engl J Med 2015;372:492−4.

[173] Warkentin TE, Elavathil LJ, Hayward CPM, Johnston MA, Russett JI, Kelton JG. The pathogenesis of venous limb gangrene associated with heparin-induced thrombocytopenia. Ann Intern Med 1997;127:804−12.

[174] Warkentin TE. Should vitamin K be administered when HIT is diagnosed after administration of coumarin? J Thromb Haemost 2006;4:894—6.

[175] White PW, Sadd JR, Nensel RE. Thrombotic complications of heparin therapy: including
six cases of heparin-induced skin necrosis. Ann Surg 1979;190:595—608.

[176] Stricker H, Lammle B, Furlan M, Sulzer I. Heparin-dependent in vitro aggregation of
normal platelets by plasma of a patient with heparin-induced skin necrosis: specific diagnostic test for a rare side effect. Am J Med 1988;85:721—4.

[177] Yazbek MA, Velho P, Nadruz W, Mahayri N, Appenzeller S, Costallat LT. Extensive
skin necrosis induced by low-molecular-weight heparin in a patient with systemic lupus
erythematosus and antiphospholipid syndrome. J Clin Rheumatol 2012;18:196—8.

[178] Schindewolf M, Kroll H, Ackermann H, Garbaraviciene J, Kaufmann R, Boehncke W-H,
et al. Heparin-induced non-necrotizing skin lesions: rarely associated with heparin-
induced thrombocytopenia. J Thromb Haemost 2010;8:1486—91.

[179] Warkentin TE, Linkins L-A. Non-necrotizing heparin-induced skin lesions and the 4T's
score. J Thromb Haemost 2010;8:1483—5.

[180] Fried M, Kahanovich S, Dagan R. Enoxaparin-induced skin necrosis. Ann Intern Med
1996;125:521—2.

[181] Hume M, Smith-Petersen M, Fremont-Smith P. Sensitivity to intrafat heparin. The
Lancet 1974;261 i:.

[182] O'Toole RD. Heparin: adverse reaction. Ann Intern Med 1973;79:759—60.

[183] Cuker A, Gimotty PA, Crowther MA, Warkentin TE. Predictive value of the 4Ts scoring
system for heparin-induced thrombocytopenia: a systematic review and meta-analysis.
Blood 2012;120:4160—7.

[184] Linkins LA, Bates SM, Lee AY, Heddle NM, Wang G, Warkentin TE. Combination of
4Ts score and PF4/H-PaGIA for diagnosis and management of heparin-induced thrombocytopenia: prospective cohort study. Blood 2015;126:597—603.

[185] Cuker A, Arepally G, Crowther MA, Rice L, Datko F, Hook K, et al. The HIT Expert
Probability (HEP) Score: a novel pre -test probability model for heparin-induced thrombocytopenia based on broad expert opinion. J Thromb Haemost 2010;8:2642—50.

[186] Pishko AM, Fardin S, Lefler DS, Paydary K, Vega R, Arepally GM, et al. Prospective
comparison of the HEP score and 4Ts score for the diagnosis of heparin-induced thrombocytopenia. Blood Adv 2018;2:3155—62.

[187] Cuker A, Arepally GM, Chong BH, Cines DB, Greinacher A, Gruel Y, et al. American
Society of Hematology 2018 guidelines for management of venous thromboembolism:
heparin-induced thrombocytopenia. Blood Adv 2018;2:3360—92.

[188] Fratantoni JC, Pollet R, Gralnick HR. Heparin-induced thrombocytopenia: confirmation
of diagnosis with in vitro methods. Blood 1975;45:395—401.

[189] Sheridan D, Carter C, Kelton JG. A diagnostic test for heparin-induced thrombocytopenia. Blood 1986;67:27—30.

[190] Padmanabhan A, Jones CG, Curtis BR, Bougie DW, Sullivan MJ, Peswani N, et al. A
novel PF4-dependent platelet activation assay identifies patients likely to have heparin-
induced thrombocytopenia/thrombosis. Chest 2016;150:506—15.

[191] Amiral J, Bridey F, Wolf M, Boyer-Neumann C, Fressinaud E, Vissac AM, et al.
Antibodies to macromolecular platelet factor 4-heparin complexes in heparin-induced
thrombocytopenia: a study of 44 cases. Thromb Haemost 1995;73:21—8.

[192] Warkentin TE, Sheppard JI, Moore JC, Sigouin CS, Kelton JG. Quantitative interpreta-
tion of optical density measurements using PF4-dependent enzyme-immunoassays. J
Thromb Haemost 2008;6:1304—12.

[193]   Bankova A, Andres Y, Horn MP, Alberio L, Nagler M. Rapid immunoassays for diagnosis of heparin-induced thrombocytopenia: comparison of diagnostic accuracy, reproducibility, and costs in clinical practice. PLoS One 2017;12:e0178289.

[194]   Whitlatch NL, Kong DF, Metjian AD, Arepally GM, Ortel TL. Validation of the high-dose heparin confirmatory step for the diagnosis of heparin-induced thrombocytopenia. Blood 2010;116:1761−6.

[195]   Francis JL. Detection and significance of heparin-platelet factor 4 antibodies. Semin Hematol 2005;42:S9−14.

[196]   Warkentin TE, Sheppard J-AI, Linkins L-A, Arnold DM, Nazy I. High sensitivity and specificity of an automated IgG-specific chemiluminescence immunoassay for diagnosis of HIT. Blood 2018;132:1345−8.

[197]   Nazi I, Arnold DM, Smith JW, Horsewood P, Moore JC, Warkentin TE, et al. FcγRIIa proteolysis as a diagnostic biomarker for heparin-induced thrombocytopenia. J Thromb Haemost 2013;11:1146−53.

[198]   Schenk S, El-Banayosy A, Morshuis M, Arusoglu L, Eichler P, Lubenow N, et al. IgG classification of anti−PF4/heparin antibodies to identify patients with heparin-induced thrombocytopenia during mechanical circulatory support. J Thromb Haemost 2007;5:235−41.

[199]   Warkentin TE. PF4-dependent immunoassays and inferential detection of HIT antibodies. J Thromb Haemost 2007;5:232−4.

[200]   De Larranaga G, Martinuzzo M, Bocassi A, Fressart MM, Forastiero R. Heparin-platelet factor 4 induced antibodies in patients with either autoimmune or alloimmune antiphospholipid antibodies. Thromb Haemost 2002;88:371−3.

[201]   Warkentin TE, Heddle NM. Laboratory diagnosis of immune heparin-induced thrombocytopenia. Curr Hematol Rep 2003;2:148−57.

[202]   Lillo-Le Louet A, Boutouyrie P, Alhenc-Gelas M, Le Beller C, Gautier I, Aiach M, et al. Diagnostic score for heparin-induced thrombocytopenia after cardiopulmonary bypass. J Thromb Haemost 2004;2:1882−8.

[203]   Heeger PS, Backstrom JT. Heparin flushes and thrombocytopenia. Ann Intern Med 1986;105:143.

[204]   Matsuo T, Kario K, Chikahira Y, Nakao K, Yamada T. Treatment of heparin-induced thrombocytopenia by use of argatroban, a synthetic thrombin inhibitor. Br J Haematol 1992;82:627−9.

[205]   Lewis BE, Wallis DE, Berkowitz SD, Matthai WH, Fareed J, Walenga JM, et al. Argatroban anticoagulant therapy in patients with heparin-induced thrombocytopenia. Circulation 2001;103:1838−43.

[206]   Bartholomew JR, Prats J. Chapter 15. Bivalirudin for the treatment of heparin-induced thrombocytopenia. In: Warkentin TE, Greinacher A, editors. Heparin-induced thrombocytopenia. 5th ed. Boca Raton: CRC Press; 2013. p. 429−65.

[207]   Joseph L, Casanegra AI, Dhariwal M, Smith MA, Raju MG, Militello MA, et al. Bivalirudin for the treatment of patients with confirmed or suspected heparin-induced thrombocytopenia. J Thromb Haemost 2014;12:1044−53.

[208]   Chong BH, Magnani HN. Chapter 16. Danaparoid for the treatment of heparin-induced thrombocytopenia. In: Warkentin TE, Greinacher A, editors. Heparin-induced thrombocytopenia. 5th ed. Boca Raton: CRC Press; 2013. p. 466−88.

[209]   Chong BH, Ismail F, Cade J, Gallus AS, Gordon S, Chesterman CN. Heparin-induced thrombocytopenia: studies with a new low molecular weight heparinoid, Org 10172. Blood 1989;73:1592−6.

[210] Krauel K, Fürll B, Warkentin TE, Weitschies W, Kohlmann T, Sheppard JI, et al. Heparin-induced thrombocytopenia--therapeutic concentrations of danaparoid, unlike fondaparinux and direct thrombin inhibitors, inhibit formation of platelet factor 4-heparin complexes. J Thromb Haemost 2008;6:2160−7.

[211] Harenberg J, Zimmermann R, Schwarz F, Kubler W. Treatment of heparin-induced thrombocytopenia with thrombosis by new heparinoid. The Lancet 1983;i:986−7.

[212] Ortel TL, Gockerman JP, Califf RM, McCann RL, O'Connor CM, Metzler DM, et al. Parenteral anticoagulation with the heparinoid Lomoparan (Org 10172) in patients with heparin induced thrombocytopenia and thrombosis. Thromb Haemost 1992;67:292−6.

[213] Magnani HN. Heparin-induced thrombocytopenia (HIT): an overview of 230 patients treated with orgaran (Org 10172). Thromb Haemost 1993;70:554−61.

[214] Tardy B, Tardy-Poncet B, Viallon A, Piot M, Mazet E. Fatal danaparoid-sodium induced thrombocytopenia and arterial thromboses. Thromb Haemost 1998;80:530.

[215] Greinacher A. Chapter 14. Recombinant hirudin for the treatment of heparin-induced thrombocytopenia. In: Warkentin TE, Greinacher A, editors. Heparin-induced thrombocytopenia. 5th ed. Boca Raton: CRC Press; 2013. p. 388−428.

[216] Greinacher A, Völpel H, Janssens U, Hach-Wunderle V, Kemkes-Matthes B, Eichler P, et al. Recombinant hirudin (lepirudin) provides safe and effective anticoagulation in patients with heparin-induced thrombocytopenia: a prospective study. Circulation 1999;99:73−80.

[217] Greinacher A, Janssens U, Berg G, Böck M, Kwasny H, Kemkes-Matthes B, et al. Lepirudin (recombinant hirudin) for parenteral anticoagulation in patients with heparin-induced thrombocytopenia. Heparin-Associated Thrombocytopenia Study (HAT) investigators. Circulation 1999;100:587−93.

[218] Lubenow N, Eichler P, Lietz T, Greinacher for the HIT investigators group. Lepirudin in patients with heparin-induced thrombocytopenia-results of the third prospective study (HAT-3) and a combined analysis of HAT-1, HAT-2, and HAT-3. J Thromb Haemost 2005;3:2428−36.

[219] Tardy B, Lecompte T, Boelhen F, Tardy-Poncet B, Elalamy I, Morange P, et al. Predictive factors for thrombosis and major bleeding in an observational study in 181 patients with heparin-induced thrombocytopenia treated with lepirudin. Blood 2006;108:1492−6.

[220] Potzsch B, Hund S, Madlener K, Unkrig C, Müller-Berghaus G. Monitoring of r-hirudin anticoagulant during cardiopulmonary bypass-assessment of the whole blood ecarin clotting time. Thromb Haemost 1997;77:920−5.

[221] Tschudi M, Lammle B, Alberio L. Dosing lepirudin in patients with heparin-induced thrombocytopenia and normal or impaired renal function: a single-center experience with 68 patients. Blood 2009;113:2402−9.

[222] Kelton JG, Arnold DM, Bates SM. Nonheparin anticoagulants for heparin-induced thrombocytopenia. N Engl J Med 2013;368:737−44.

[223] Greinacher A, Alban S, Dummel V, Franz G, Mueller-Eckhardt C. Characterization of the structural requirements for a carbohydrate based anticoagulant with a reduced risk of inducing the immunological type of heparin-associated thrombocytopenia. Thromb Haemost 1995;74:886−92.

[224] Elalamy I, Lecrubier C, Potevin F, Abdelouahed M, Bara L, Marie JP, et al. Absence of in vitro cross-reaction of pentasaccharide with the plasma heparin-dependent factor of twenty-five patients with heparin-associated thrombocytopenia. Thromb Haemost 1995;74:1379−87.

[225] Warkentin TE, Pai M, Sheppard JI, Schulman S, Spyropoulos AC, Eikelboom JW. Fondaparinux treatment of acute heparin-induced thrombocytopenia confirmed by the serotonin-release assay: a 30 −month, 16 patient case series. J Thromb Haemost 2011;9:2389−96.

[226] Kang M, Alahmadi M, Sawh S, Kovacs MJ, Lazo-Langner A. Fondaparinux for the treatment of suspected heparin-induced thrombocytopenia: a propensity score-matched study. Blood 2015;125:924−9.

[227] Schindewolf M, Steindl J, Beyer-Westendorf J, Schellong S, Dohmen PM, Brachmann J, et al. Use of fondaparinux off-label or approved anticoagulants for management of heparin-induced thrombocytopenia. J Am Coll Cardiol 2017;70:2636−48.

[228] Bhatt VR, Aryal MR, Armitage JO. Nonheparin anticoagulants for heparin-induced thrombocytopenia. N Engl J Med 2013;368:2333−4.

[229] Aljabri A, Huckleberry Y, Karnes J, Gharaibeh M, Kutbi HI, Raz Y, et al. Cost−effectiveness of anticoagulants for the management of suspected heparin-induced thrombocytopenia in the US. Blood 2016;128:3043−51.

[230] Krauel K, Hackbarth C, Furli B, Greinacher A. Heparin-induced thrombocytopenia: in vitro studies on the interaction of dabigatran, rivaroxaban, and low-sulfated heparin, with platelet factor 4 and anti−PF4/heparin antibodies. Blood 2012;119:1248−55.

[231] Ng HJ, Than H, Teo ECY. First experiences with the use of rivaroxaban in the treatment of heparin-induced thrombocytopenia. Thromb Res 2015;135:205−7.

[232] Warkentin TE, Pai M, Linkins LA. Direct oral anticoagulants for treatment of HIT: update of Hamilton experience and literature review. Blood 2017;130:1104−13.

[233] Linkins LA, Warkentin TE, Pai M, Shivakumar S, Manji RA, Wells PS, et al. Rivaroxaban for treatment of suspected or confirmed heparin-induced thrombocytopenia study. J Thromb Haemost 2016;14:1206−10.

[234] Frame JN, Mulvey KP, Phares JC, Anderson MJ. Correction of severe heparin-associated thrombocytopenia with intravenous immunoglobulin. Ann Intern Med 1989;111:946−7.

[235] Park BD, Kumar M, Nagalla S, De Simone N, Aster RH, Padmanabhan A, et al. Intravenous immunoglobulin as an adjunct therapy in persisting heparin-induced thrombocytopenia. Transfus Apher Sci 2018;57:561−5.

[236] Padmanabhan A, Jones CG, Pechauer SM, Curtis BR, Bougie DW, Irani MS, et al. IVIg for treatment of severe refractory heparin-induced thrombocytopenia. Chest 2017;152:478−85.

[237] Warkentin TE, Climans TH, Morin P-A. Intravenous immune globulin to prevent heparin-induced thrombocytopenia. N Engl J Med 2018;378:1845−8.

[238] Jones CG, Pechauer SM, Curtis BR, Bougie DW, Aster RH, Padmanabhan A. Normal plasma IgG inhibits HIT antibody-mediated platelet activation: implications for therapeutic plasma exchange. Blood 2018;131:703−6.

[239] Warkentin TE, Sheppard J-AI, Chu FV, Kapoor A, Crowther MA, Gangji A. Plasma exchange to remove HIT antibodies: dissociation between enzyme-immunoassay and platelet activation test reactivities. Blood 2015;125:195−8.

[240] Janson PA, Moake JL, Carpinito G. Aspirin prevents heparin-induced platelet aggregation in vivo. Br J Haematol 1983;53:166−8.

[241] Herbert J-M, Savi P, Jeske WP, Walenga JM. Effect of SR121566A, a potent GPIIb/IIIa antagonist, on the HIT serum/heparin-induced platelet mediated activation of human endothelial cells. Thromb Haemost 1998;80:326−31.

[242] Mak K-H, Kottke-Marchant K, Brooks LM, Topol EJ. In vitro efficacy of platelet glyco-protein IIb/IIIa antagonist in blocking platelet function in plasma of patients with heparin-induced thrombocytopenia. Thromb Haemost 1998;80:989−93.

[243] Reilly MP, Sinha U, André P, Taylor SM, Pak Y, Deguzman FR, et al. PRT-060318, a novel Syk inhibitor, prevents heparin-induced thrombocytopenia and thrombosis in a transgenic mouse model. Blood 2011;117:2241−6.

[244] Kizlik-Masson C, Deveuve Q, Zhou Y, Vayne C, Thibault G, McKenzie SE, et al. Cleavage of anti−PF4/heparin IgG by a bacterial protease and potential benefit in heparin-induced thrombocytopenia. Blood 2019;133:2427−35.

[245] Bircher AJ, Fluckiger R, Buchner SA. Eczematous infiltrated plaques to subcutaneous heparin: a type IV allergic reaction. Br J Dermatol 1990;123:507−14.

[246] Schindewolf M, Schwaner S, Wolter M, Kroll H, Recke A, Kaufmann R, et al. Incidence and causes of heparin-induced skin lesions. CMAJ 2009;181:477−81.

[247] Bank I, Libourel EJ, Middeldorp S, Van der Meer J, Buller HR. High rate of skin com-plications due to low-molecular-weight heparins in pregnant women. J Thromb Haemost 2003;1:859−61.

[248] Schindewolf M, Gobst C, Kroll H, Recke A, Louwen F, Wolter M, et al. High incidence of heparin-induced allergic delayed-type hypersensitivity reactions in pregnancy. J Allergy Clin Immunol 2013;132:131−9.

[249] Wutschert R, Piletta P, Bounameaux H. Adverse skin reactions to low molecular weight heparins: frequency, management and prevention. Drug Saf 1999;20:515−25.

[250] Odeh M, Oliven A. Urticaria and angioedema induced by low-molecular weight heparin. The Lancet 1992;340:972−3.

[251] Bircher AJ, Itin PH, Buchner SA. Skin lesions, hypereosinophilia, and subcutaneous hep-arin. The Lancet 1994;343:861.

[252] Ludwig RJ, Schindewolf M, Alban S, Kaufmann R, Lindhoff-Last E, Boehncke WH. Molecular weight determines the frequency of delayed type hypersensitivity reactions to heparin and synthetic oligosaccharides. Thromb Haemost 2005;94:1265−9.

[253] Grassegger A, Fritsch P, Reider N. Delayed-type hypersensitivity and cross-reactivity to heparins and danaparoid: a prospective study. Dermatol Surg 2001;27:47−52.

[254] Gaigl Z, Pfeuffer P, Raith P, Bröcker EB, Trautmann A. Tolerance to intravenous hepa-rin in patients with delayed-type hypersensitivity to heparins: a prospective study. Br J Haematol 2005;128:389−92.

[255] Hall JC, McConahay D, Gibson D, Crockett J, Conn R. Heparin necrosis. An anticoagu-lation syndrome. JAMA 1980;244:1831−2.

[256] Balestra B, Quadri P, Dermarmels Biasiutti F, Furlan M, Lämmle B. Low molecular weight heparin-induced thrombocytopenia and skin necrosis distant from injection sites. Eur J Haematol 1994;53:61−3.

[257] Handschin AE, Trentz O, Kock HJ, Wanner GA. Low molecular weight heparin-induced skin necrosis-a systematic review. Langenbecks Arch Surg 2005;390:249−54.

[258] Warkentin TE, Roberts RS, Hirsh J, Kelton JG. Heparin-induced skin lesions and other unusual sequelae of the heparin-induced thrombocytopenia syndrome: a nested cohort study. Chest 2005;127:1857−61.

[259] Apsner R, Horl WH, Sunder-Plassmann G. Dalteparin-induced alopecia in hemodialysis patients: reversal by regional citrate anticoagulation. Blood 2001;97:2914−15.

[260] Barnes C, Deidun D, Hynes K, Monagle P. Alopecia and dalteparin: a previously unre-ported association. Blood 2000;96:1618−19.

[261] Wang Y-Y, Po HL. Enoxaparin-induced alopecia in patients with cerebral venous thrombosis. J Clin Pharm Therap 2006;31:513—17.

[262] Tosi A, Misciali C, Piraccini BM, Peluso AM, Bardazzi F. Drug-induced hair loss and hair growth. Incidence, management and avoidance. Drug Saf 1994;10:310—17.

[263] Griffith GC, Nichols Jr G, Asher JD, Flanagan B. Heparin osteoporosis. JAMA 1965;193:85—8.

[264] Rajgopal R, Bear M, Butcher MR, Shaughnessy SG. The effects of heparin and low molecular weight heparins on bone. Thromb Res 2008;122:293—8.

[265] Shaughnessy SG, Young E, Deschamps P, Hirsh J. The effects of low molecular weight and standard heparin on calcium loss from fetal rat calvaria. Blood 1995;86:1368—73.

[266] Muir JM, Hirsh J, Weitz JI, Andrew M, Young E, Shaughnessy SG. A histomorphometric comparison of the effects of heparin and low-molecular-weight heparin on cancellous bone in rats. Blood 1997;89:3236—42.

[267] Bhandari M, Hirsh J, Weitz JI, Young E, Venner TJ, Shaughnessy SG. The effects of standard and low molecular weight heparin on bone nodule formation in vitro. Thromb Haemost 1998;80:413—17.

[268] Kock HJ, Handschin AE. Inhibition of human osteoblast growth by unfractionated heparin and by low molecular weight heparins in vitro. Thromb Haemost 2002;88:361—2.

[269] Shaughnessy SG, Hirsh J, Bhandari M, Muir JM, Young E, Weitz JI. A histomorphometric evaluation of heparin-induced bone loss after discontinuation of heparin treatment in rats. Blood 1999;93:1231—6.

[270] Barbour LA, Kick SD, Steiner JF, LoVerde ME, Heddleston LN, Lear JL, et al. A prospective study of heparin-induced osteoporosis in pregnancy using bone densitometry. Am J Obstet Gynecol 1994;170:862—9.

[271] Douketis JD, Ginsberg JS, Burows RF, Duku EK, Webber CE, Brill-Edwards P. The effects of long-term heparin therapy during pregnancy on bone density. Thromb Haemost 1996;75:254—7.

[272] Oster JR, Singer I, Fishman LM. Heparin-induced aldosterone suppression and hyperkalemia. Am J Med 1995;98:575—86.

[273] Dukes GE, Sanders SW, Russo Jr J, Swenson E, Burnakis TG, Saffle JR, et al. Transaminase elevations in patients receiving bovine or porcine heparin. Ann Intern Med 1984;100:646—50.

[274] Monreal M, Lafoz E, Salvador R, Roncales J, Navarro A. Adverse effects of three different forms of heparin therapy: thrombocytopenia, increased transaminases, and hyperkalaemia. Eur J Clin Pharmacol 1989;37:415—18.

Index

Printed in the United States
By Bookmasters